Electrochemical Detectors
Fundamental Aspects and Analytical Applications

Electrochemical Detectors
Fundamental Aspects and Analytical Applications

Edited by
T. H. Ryan

EDT Research
London, England

PLENUM PRESS • NEW YORK AND LONDON

Library of Congress Cataloging in Publication Data

Main entry under title:

Electrochemical detectors.

"Proceedings of a symposium sponsored by the Analytical and Faraday
Division of the Royal Society of Chemistry, held September 15–16, 1981, in
Chelsea College, London, England."
 Includes bibliographical references and index.
 1. Chemical detectors—Congresses. 2. Electrochemical apparatus—
Congresses. I. Ryan, T. H. II. Royal Society of Chemistry (Great Britain).
Analytical and Faraday Division.
TP159.C46E43   1984   543'.0894   84-11475
ISBN-13: 978-1-4615-9399-7   e-ISBN-13: 978-1-4615-9397-3
DOI: 10.1007/978-1-4615-9397-3

Proceedings of a symposium sponsored by the
Analytical and Faraday Division of the Royal Society of Chemistry,
held September 15–16, 1981,
in Chelsea College, London, England

©1984 Plenum Press, New York
Softcover reprint of the hardcover 1st edition 1984
A Division of Plenum Publishing Corporation
233 Spring Street, New York, N.Y. 10013

PREFACE

The widely perceived utility of electrochemical detectors in High Performance Liquid Chromatography has focussed attention on a number of disparate aspects of electrochemistry related to their successful design and application. The papers in this volume deal with an extraordinarily wide range of topics but all have the common focus of electrochemical detection as a practical chromatographic tool. While it is certainly not essential to be familiar with the theoretical principles in order to utilize it successfully, the determined user of electrochemical detector will seek to have an understanding of the background. Some of the following pages will provide an excellent grounding as well as pointing the potential user in the direction of proven and possible applications in a variety of fields.

The meeting, of which this book is a record, was the fifth event in the biannual Anglo-Czech Symposia in Electrochemistry. The organizers of that meeting were extremely pleased to welcome a select group of scientists from the J Heyrovsky Institute of Physical Chemistry and Electrochemistry to the meeting, thus continuing a tradition of cooperation and friendship going back over ten years. The contributions of the visitors to the scientific content of the meeting were enthusiastically received and their participation in the informal and social activities can only have furthered the cause of cooperation and good will between our two countries.

I would like to take this opportunity to thank the authors of the contributions for preparing their manuscripts, especially as the decision to publish them was made only after the meeting had taken place. I am indebted to the Editor and Publishers of "Surface Technology" for permission to print one of the papers (by Douglas A Richards) which has already appeared therein. Abstracts of the Czech contributions were also published in the same volume.

Finally my thanks must go to Dr Robert Andrews of Plenum Press who was able to arrange preparation of the camera-ready copy and who was able to guide a first-time editor in publishing matters.

T H Ryan

CONTENTS

NEW ELECTROANALYTICAL TECHNIQUES APPLIED TO

MEDICINE AND BIOLOGY

W. John Albery and Barry G. D. Haggett

Department of Chemistry
Imperial College
London SW7 2AY

INTRODUCTION

In this article we shall describe the development of new electroanalytical techniques for the determination of three different classes of compounds of interest to biologists and clinicians. We shall first describe a novel detector using a ring-disc electrode[1] for the estimation and identification of proteins, in which the protein is titrated with bromine electrogenerated on the disc electrode and the unreacted bromine is measured on the ring electrode. The method is a general one for all proteins and amino acids and can be realised using a wall jet cell [2,3]; this means that the detector can easily be coupled to a chromatographic column. Secondly we shall describe the joint work we have been carrying out with Dr Clive Hahn of the Radcliffe Infirmary in Oxford on the development of sensors for the gases used in anaesthesia. These sensors are based on the Clark membrane electrode for oxygen [4]. We have now constructed and tested sensors for N_2O[5,6] halothane[7] and CO_2 [8]. The response time of the CO_2 electrode is short enough for breath by breath analysis. Finally we shall describe work using microelectrodes for the in vivo measurement of catecholamines, their metabolites and ascorbic acid in the brains of freely moving rats. This program of work is carried out in collaboration with Dr M. Fillenz of the Department of Physiology in Oxford.

Diffusion Layer Titration of Proteins

The principle of the ring-disc titration curve[1] is to generate Br_2 on the disc electrode, which then reacts with the target molecules in the solution; the unreacted Br_2 is measured down-stream on the ring electrode:

Disc Electrode .... $Br^- \longrightarrow \frac{1}{2} Br_2 + e$

Solution .... $n\ Br_2 + P \xrightarrow{k}$ Products

Ring Electrode .... $\frac{1}{2} Br_2 + e \longrightarrow Br^-$

The ring electrode potential is set so that all the Br_2 which reaches it is reduced. The rate constant k for the titration reaction has to be greater than about $10^5\ dm^3\ mol^{-1}\ s^{-1}$ for the reaction to intercept the Br_2 in its passage from disc to ring. At a pH of 9.2, where the "bromine" is present as OBr^- rather than Br_2, typical reactions which consume Br_2 are

$$-\underset{|}{\overset{|}{C}} - NH_2 + 2OBr^- \longrightarrow -\underset{|}{\overset{|}{C}} - NBr_2 + 2OH^-$$

$$-CH_2 - S - S - CH_2^- + 5OBr^- + 2OH^- \longrightarrow 2(-CH_2 - SO_3^-) + 5Br^- + H_2O$$

$$-\langle\!\!\bigcirc\!\!\rangle - OH + 2OBr^- \longrightarrow -\langle\!\!\bigcirc\!\!\rangle \overset{Br}{\underset{Br}{-}} OH + 2OH^-$$

$$-C = C - + Br^- + OBr^- + H_2O \longrightarrow -\underset{Br}{C} - \underset{Br}{C} - + 2OH^-$$

We have found that at this basic pH a single molecule of protein will consume several hundred molecules of OBr^-. We call this number the bromine number of the protein. This ability of OBr^- to probe the protein for the brominatable groups is an important advantage of titration technique as compared to direct determination by reduction or oxidation of the protein on the electrode.

A typical rotating ring-disc titration curve is shown in Figure 1 together with a strip cartoon showing the bromine zone spreading out from the disc electrode as the generating disc current is increased. Inside the bromine zone the protein is fully brominated; outside it there is no free bromine because it would be rapidly gobbled up by the unbrominated protein. The ring current measures the amount of Br_2 reaching the ring electrode. The broken straight line shows the collection efficiency that would be observed if there was no consumption of Br_2 by the titration reaction. The displacement of the ring current from this line is proportional to the concentration of protein. Typical results are shown in Figure 2. We have shown that up to 24 different proteins can be titrated in this way. Typical results for bromine numbers at pH 9.2 are collected in Table 1.

At a pH of 5, some of the bromination reactions listed above take place so slowly that they do not take place on the time scale of the ring-disc titration. In particular protonation of the NH_2 groups prevents their bromination. Hence the bromine number measured at pH

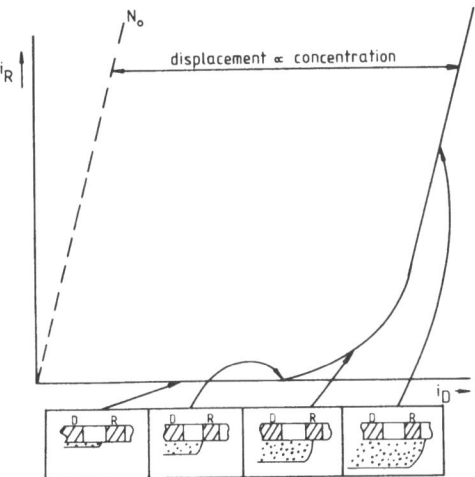

Fig. 1. Ring-disc titration curve showing the ring current as a function of the disc current. The insets show the bromine zone spreading out from the disc towards the ring electrode.

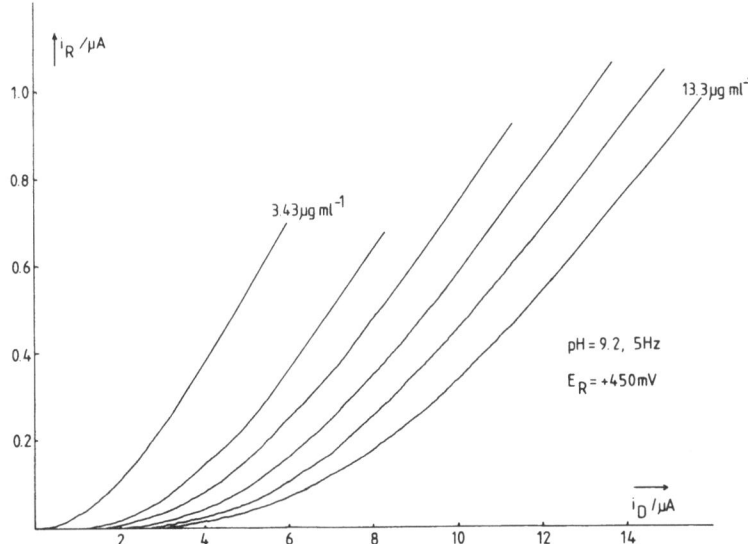

Fig. 2. Typical titration curves for the bromination of haemoglobin showing the effect of the successive addition of aliquots of the protein.

5 is always less than that at 9.2. We have shown that the ratio of the bromine numbers is a characteristic of the particular protein and hence can be used to help identify the protein. Some results are given in Table 2. We can also combine the bromine number ratio with

Table 1. Bromine Numbers at pH 9.2

Protein	n_{Br} (pH 9.2)
B S A	1.099 ± .026
M S A	1.100 ± .022
Thyroglobulin	.399 ± .010
Haemoglobin	1.417 ± .066
Cytochrome c	2.195 ± .038
γ–Amylase	.671 ± .013
Trypsinogen	1.217 ± .016
Transferrin	.988 ± .029
β–Lactoglobulin	1.150 ± .009
γ–Chymotrypsinogen	.975 ± .010
Conalbumin I	.953 ± .015
Conalbumin II	.753 ± .007
Catalase	1.163 ± .019
Myoglobin (whale)	1.982 ± .021
Myoglobin (horse)	1.783 ± .028

the measurement of optical density at λ = 280 nm using a simple flow through u.v. detector to form a second ratio which is characteristic of the protein. Using the two ratios a diagram can be constructed, as shown in Figure 3, in which common proteins are separated; from its position on the diagram an unknown protein can be identified.

Although hitherto the ring-disc titrations have been performed using a rotating electrode, we have developed the wall-jet ring-disc electrode. In the wall-jet electrode (see Figure 4) a jet of solution impinges on the centre of a disc electrode and then spreads out in a radial direction [2,3]. This flow carries material from the disc electrode to the ring electrode. We have shown[9] that there is a close connection between the theory of the rotating ring-disc electrode and the wall-jet ring-disc electrode. The theoretical expressions for the collection efficiency and for the diffusion layer titration curve for the rotating system can be used for the wall-jet system providing that the geometric parameters α, β and β_J are defined, as in Table 3; α describes the geometry of the gap, β the geometry of the ring electrode and β_J the location of the edge of the bromine zone (see Figure 1) when it is on the ring electrode. The same expressions also hold for collection efficiency and for titration curves for double channel or double tube electrodes and the appropriate definitions for α, β and β_J are also given in Table 3. We have verified that theory and experiment for the wall-jet ring-disc titration curves are in reasonable agreement and we have shown that successful diffusion layer titrations of some 6 different proteins can be carried out using the wall-jet cell.

Table 2. Ratio of Bromine Numbers at pH 9.2 and pH 5

Protein	n_{Br} (9.2)$/n_{Br}$ (5)
B S A	9.24 ± .38
M S A	8.80 ± .27
Thyroglobulin	3.73 ± .14
Haemoglobin	3.67 ± .20
Cytochrome c	5.92 ± .26
γ-Amylase	3.26 ± .19
Trypsinogen	6.91 ± .48
Transferrin	4.73 ± .19
β-Lactoglobulin	5.43 ± .11
γ-Chymotrypsinogen	2.89 ± .11
Conalbumin I	4.81 ± .16
Conalbumin II	3.54 ± .08
Catalase	3.08 ± .08
Myoglobin (whale)	1.30 ± .04
Myoglobin (horse)	-

Sensors for Clinical Gases

Membrane electrodes for the determination of O_2 are commonplace. In our joint work with Dr Hahn we have developed membrane electrodes for the determination of N_2O[5,6] and of halothane (CHClBrCF$_3$)[7] which are the two most important gases used in anaesthesia. The construction of the electrodes is similar to that of a convential membrane electrode. We showed that N_2O is reduced quantitatively on a silver cathode [6]. Typical current voltage curves are shown in Figure 5 together with the Levich plot of the limiting currents. Figure 6 shows the typical response of a membrane electrode to mixtures of O_2 and N_2O. It is clear that by choice of polarizing potential the same silver electrode can be used to determine both O_2 and N_2O [5]. We have also used a silver catheter electrode as a continuous on line monitor of O_2 and N_2O concentrations in the arterial blood of an anaesthetized dog [10]. The electrochemical measurements were compared with those from an on line mass spectrometer; results are presented in Figures 7 and 8. Good straight lines are obtained; the cost of instrumentation for the y axis is less than one per cent of the cost for the x axis.

In this work we found that halothane interfered with the oxygen currents and we have since shown that halothane can also be reduced quantitatively on a silver cathode [7]. Typical results on the rotating disc electrode are shown in Figure 9. We have shown that the overall reaction is

$$2e + H_2O + CHClBrCF_3 \xrightarrow{\text{Ag}} Br^- + OH^- + CH_2ClCF_3$$

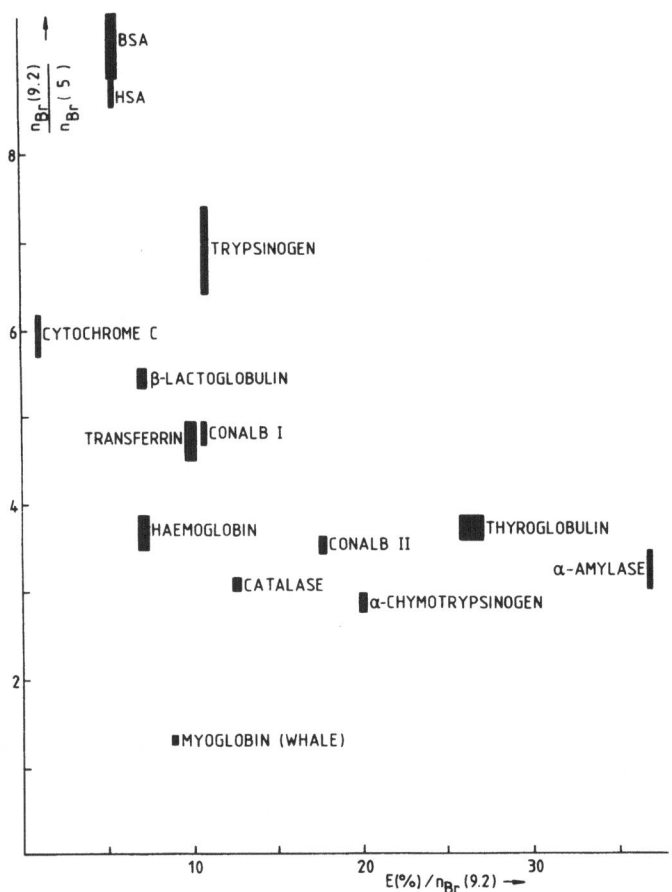

Fig. 3. Diagram showing how proteins can be identified by the ratio
of their bromine numbers at pH 9.2 and at pH 5 and the ratio
of the absorbance at λ = 280 nm and the bromine number at
pH 9.2.

Results for the performance of a halothane membrane electrode
are given in Figure 10. The membrane has to be permeable to halothane
and we have used silicone rubber or silastic membranes. The half
wave potential for the reduction of halothane is close to that of
oxygen and therefore it is difficult to separate the waves on the
silver electrode. However by constructing two working electrodes –
one of silver on which halothane reacts and one of platinum on which
it does not – and by placing both electrodes behind the same membrane
one can determine both O_2 and halothane.

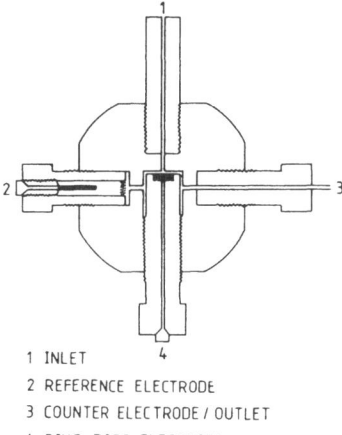

1 INLET
2 REFERENCE ELECTRODE
3 COUNTER ELECTRODE / OUTLET
4 RING-DISC ELECTRODE

Fig. 4. Typical wall-jet ring-disc electrode.

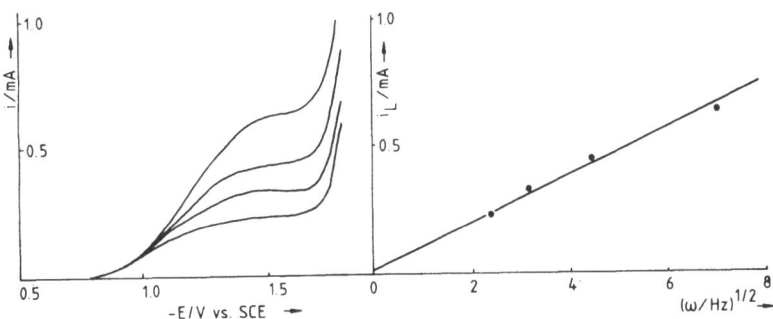

Fig. 5. Current voltage curves for the reduction of N_2O on a silver
         rotating disc electrode. The limiting currents obey the
         Levich equation.

Next we turn to the measurement of CO_2. The Severinghaus membrane electrode[11] relies on the effect of the CO_2/HCO_3^- equilibrium
on the pH of the electrolyte solution. The electrode suffers from
interference from other acidic or basic gases and has a longish
response time because the sensor is a glass electrode. We have
constructed a membrane electrode for the determination of CO_2 which
works on the amperometric principle [8]. CO_2 is reduced quantitatively in dimethylsulphoxide (DMSO). Figure 11 shows current voltage
curves and the Levich plot for reducing CO_2 on a silver electrode.
In damp DMSO we find that two electrons are consumed and ring-disc
studies show that the principal product is formate ion:

$$CO_2 + H_2O + 2e \longrightarrow HCOO^- + OH^-$$

Table 3.   Definitions of α, β and β_J for Double Electrodes

Electrode	α [a]	β [b]	β_J [c]
Rotating Ring-disc	[d] $(r_2/r_1)^3-1$	$(r_3/r_1)^3-(r_2/r_1)^3$	$(r_J/r_1)^3-(r_2/r_1)^3$
Wall-jet Ring-disc	[d] $(r_2/r_1)^{9/8}-1$	$(r_3/r_1)^{9/8}-(r_2/r_1)^{9/8}$	$(r_J/r_1)^{9\ 8}-(r_2/r_1)^{9/8}$
Double Channel Electrode	[e] $(\ell_2/\ell_1)-1$	$(\ell_3/\ell_1)-(\ell_2/\ell_1)$	$(\ell_J/\ell_1)-(\ell_2/\ell_1)$

Notes
a) α describes the geometry of the gap.
b) β describes the geometry of the downstream electrode.
c) β_J describes the edge of the Br_2 zone on the downstream
        electrode in a diffusion layer titration.
d)   Radius of disc is r_1; inner and outer radii of ring are r_2
     and r_3 respectively.
e)   Measuring always from the upstream edge of upstream
     electrode the gap lies between ℓ_1 and ℓ_2 while the
     downstream edge of the downstream electrode is at ℓ_3.

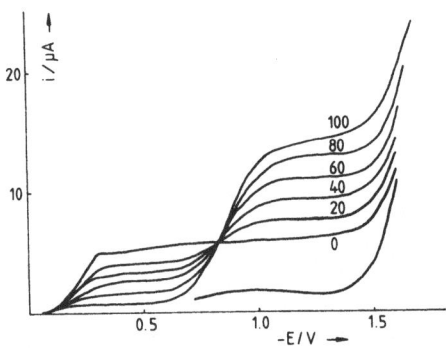

Fig. 6.   Response of a membrane electrode with silver cathode to
          mixtures of O_2 and N_2O. Each current voltage curve is
          labelled with the percentage of N_2O in the mixture.

Figure 12 shows the linear response for a membrane electrode. The
response time of this membrane electrode is less than a second and
Figure 13 shows breath by breath response for both O_2 and CO_2. These
measurements were obtained with two working electrodes, one made of

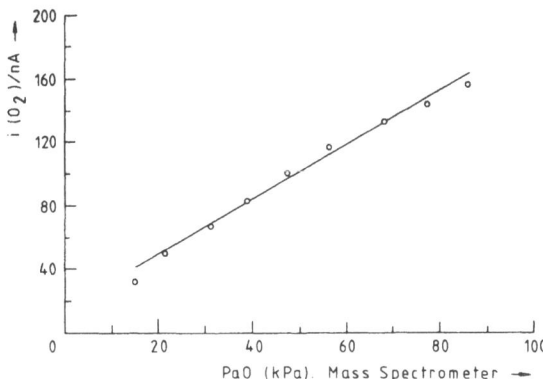

Fig. 7. Comparison of current response from an <u>in vivo</u> catheter
electrode with the determination of PO_2 from mass
spectrometer.

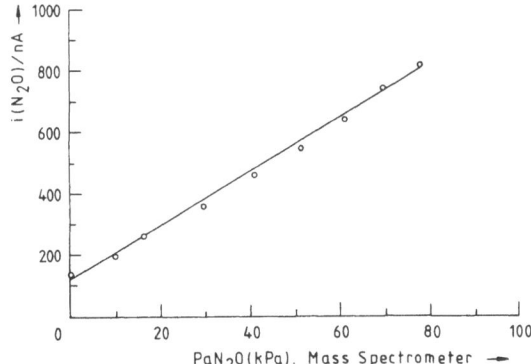

Fig. 8. Comparison of current response from an <u>in vivo</u> catheter
electrode with the determination of PN_2O from mass
spectrometer.

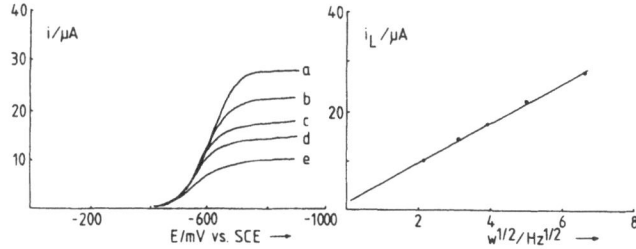

Fig. 9. Current voltage curves for the reduction of halothane on a
silver rotating disc electrode. The limiting currents obey
the Levich equation.

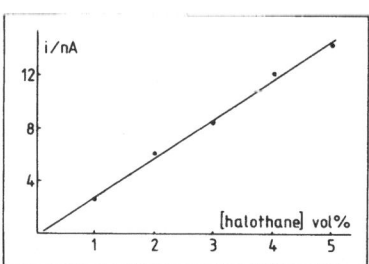

Fig. 10. Response of membrane electrode to concentration of halothane.

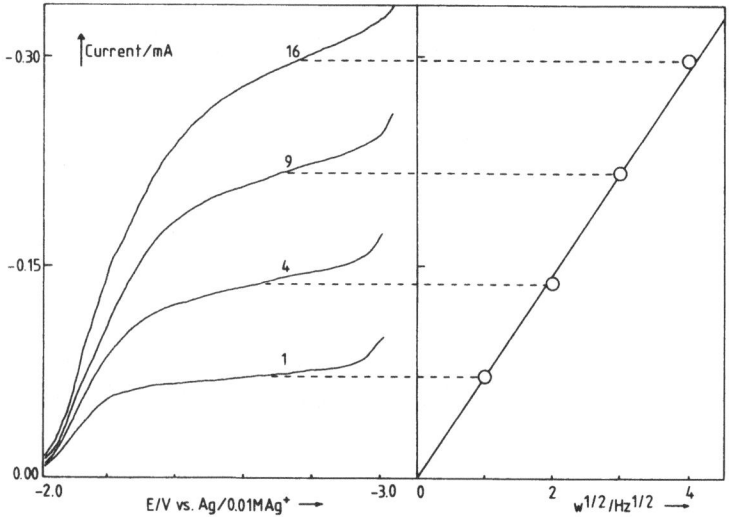

Fig. 11. Current voltage curves for the reduction of CO_2 in DMSO. The limiting currents obey the Levich equation.

gold for reducing O_2 and the other of silver for reducing both CO_2 and O_2 behind the same membrane. The CO_2 concentration was measured as the difference between the currents from the two electrodes. The ability to measure both CO_2 and O_2 breath by breath opens up new possibilities in clinical practice and in respiration research.

For the measurement of lower levels of CO_2 it is necessary to remove the much larger current from the reduction of ambient oxygen. We have shown that one can use a metallised membrane as an efficient electrochemical filter [8]. The principle of the method is shown in Figure 14. Our metallized membranes were homemade by the method of Bruckenstein[12] rather than by the earlier technique pioneered by Bergmann [13]. In the Bruckenstein method gold resinate solution is painted into a porous teflon membrane and fired with a hot air gun.

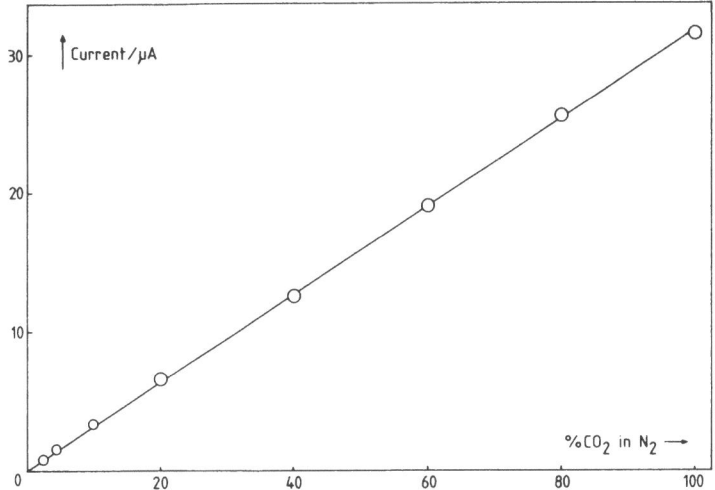

Fig. 12.   Response of membrane electrode to concentration of CO_2.

Scanning electron microscopy shows that the resultant gold clusters are deeply embedded in the teflon matrix and this leads to an efficient filter. Figure 15 shows some typical results for the removal of O_2 on a gold metallized membrane. When the filter is switched on the current for the reduction of O_2 on the inner working electrode is less than one per cent of its original value. Furthermore the current on the membrane electrode is proportional to the oxygen concentration. Sandwich electrodes with a number of metallized membranes therefore provide new and exciting possibilities for the construction of a single sensor for analysing mixtures of gases.

In Vivo Studies of Brain Chemistry

The use of implantable microelectrodes to study brain chemistry in vivo was pioneered by Adams[14,15] and by Lane and Hubbard [16,17]. It has been shown that neurotransmitters and their metabolites, such as noradrenaline, can be oxidized on graphite electrodes. The desired electrochemical reaction[18] is

where R is $CHOHCH_2NH_2$.

Different techniques such as linear sweep voltammetry (LSV) with and without semi-differentiation (SD), pulse voltammetry and differ-

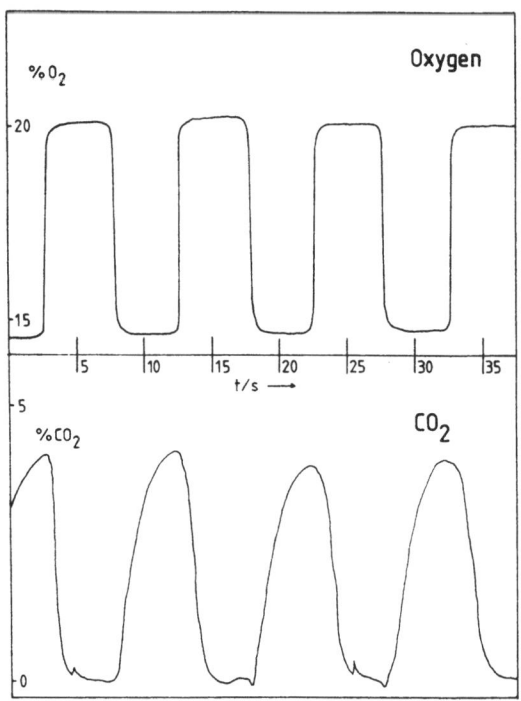

Fig. 13. Breath by breath response of membrane electrode to O_2 and
          CO_2.

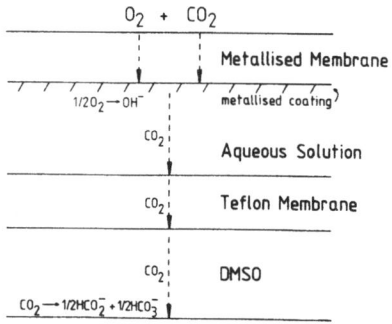

Fig. 14. The use of a metallized membrane electrode as an electro-
          chemical filter to remove O_2 so that smaller concentrations
          of CO_2 can be determined.

ential double pulse voltammetry have all been used. Theoretically
the differential double pulse method is very attractive and some of
our in vitro results are shown in Figure 16 [19]. However we have
found that when the electrodes are implanted in vivo, better results
are obtained using linear sweep voltammetry with semi-differen-

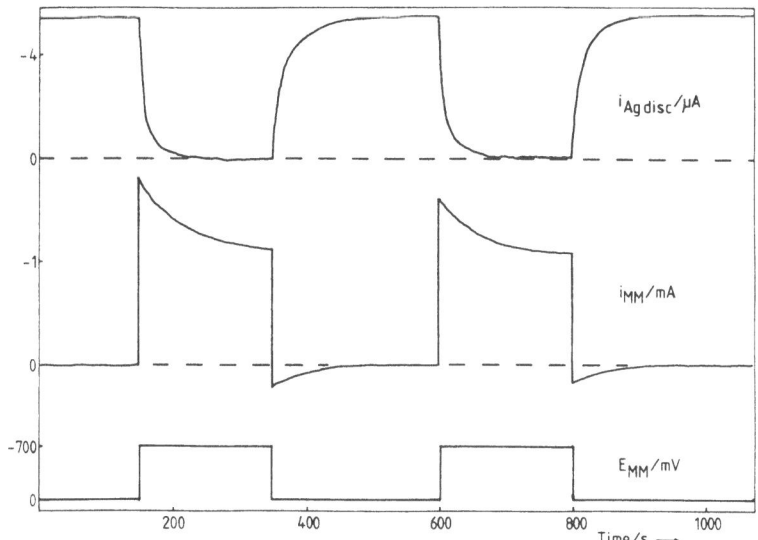

Fig. 15. The efficiency of a gold metallised membrane as a filter
for the removal of O_2. The bottom trace shows the
electrode potential of the metallized membrane, the middle
trace the current on the membrane electrode and the top
trace the current due to O_2 on the interior Ag cathode.
The filter is 99% efficient.

tiation. In our hands the signal is less noisy and better resolution
is obtained. The use of semi-differentiation[17] allows separate
current peaks from complicated mixtures to be obtained. In Figure 17
we sketch a normal cyclic voltammogram peak and the effect of semi-
differentiation. The current, after the maximum in the original
voltammogram, declines slowly according to $t^{-\frac{1}{2}}$ since at these potentials the electrode will remove all the material that reaches it,
giving the usual Cottrell behavior. This means that the rising
current of the next peak is masked by the $t^{-\frac{1}{2}}$ current of the preceding peak. From the Laplace transform of Fick's second law of
diffusion we obtain the general result

$$\left(\frac{\partial \bar{c}}{\partial x}\right)_o = \left(\frac{s}{D}\right)^{\frac{1}{2}} \left[\frac{c_\infty}{s} - \bar{c}_o\right]$$

where s is the transformed variable of t and c_∞ is the bulk concentration of the species reacting at the electrode. Semi-differentiation[20] is equivalent to multiplying by $s^{\frac{1}{2}}$ and so

$$\frac{d^{\frac{1}{2}}}{dt^{\frac{1}{2}}} \left(\frac{\partial c}{\partial x}\right)_o = \frac{1}{D^{\frac{1}{2}}} \frac{d}{dt} (c_\infty - c_o)$$

That is, the semi-differentiated current measures the differential change in suface concentration of the target species. Hence the semi-differentiated voltammogram gives a much narrower peak for each system ($\sim$ 100 mV) centred for a reversible system on the E^{θ} of the substance. The actual operation of semi-differentiation can be achieved by a simple analogue circuit [21].

As regards electrodes we have used silicone oil carbon paste electrodes made according to the method of Adams and co-workers [15]; the electrode diameter is 300 μm. We have also used carbon fiber electrodes with typical diameters of 8 μm. These electrodes have the advantage of being smaller and French workers[22] report that better resolution can be obtained in the voltammogram. On the other hand, when used in vivo, the carbon fiber electrodes poison after about eight hours, while the carbon paste electrodes can be used in the same freely moving rat over a period of several week. We therefore use carbon paste electrodes. The electrode is implanted stereotaxically into the caudate nucleus or the hippocampus of the rat. The leads come through a swivel and after implantation the rat is freely moving and does most of the things that rats like doing.

In order to be able to record and analyse data continuously, in collaboration with Dr N. J. Goddard of our Microprocessor Unit, we have developed the microprocessor controlled instrumentation shown in Figure 18 [23]. Without microprocessor instrumentation it was difficult to monitor the rats continuously and almost impossible to analyse the quantities of data spewing out on roll after roll of chart paper.

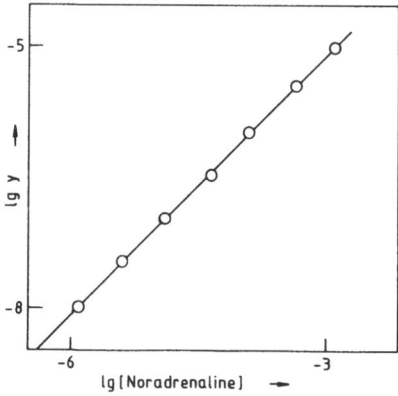

Fig. 16.   Determination of noradrenaline in vitro, using differential double pulse voltanmetry. On the y axis, y is the product of the peak height times the peak width at half-height. We have shown that this product is a more reliable measure of concentration since it takes into account the gradual poisoning of the electrode[19].

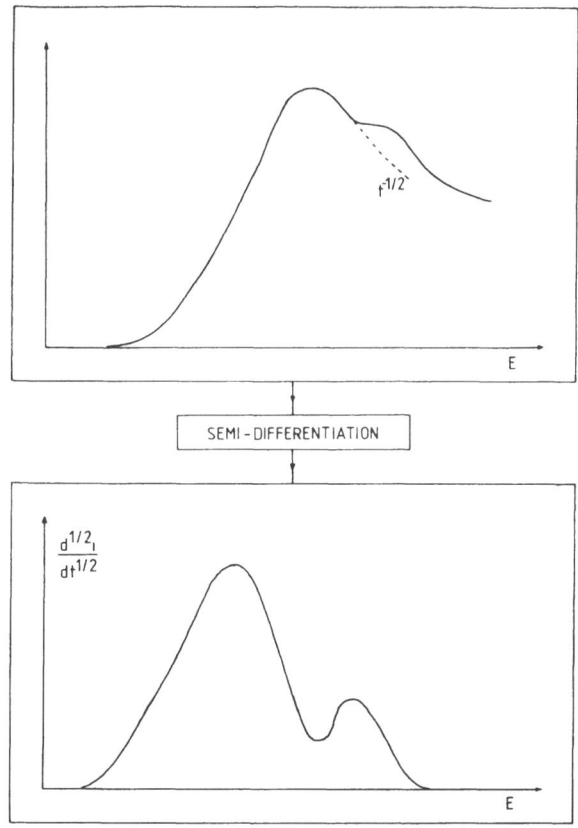

Fig. 17.   The top figure shows a voltammogram obtained by linear
           sweep. The slow fall off of the current from the peak
           $(t^{-\frac{1}{2}})$ masks the smaller peak due to a second substance.
           Semi-differentiation gives zero current in this region and
           hence the peaks are better reduced.

      Figure 19 shows a typical semi-differentiated voltammogram
obtained in the rat striatum. It can be seen that there are four
peaks; the positions of these peaks does depend on the sweep rate
which suggests that the systems are not reversible.

      Now the next question is, what substances are causing the dif-
ferent peaks? Many workers have assumed that peak 1 is caused by the
oxidation of catecholamines such as noradrenaline. However there is
a problem in that the oxidation potentials for ascorbic acid and the
catecholamines are not very different. A further problem we have
found is that oxidation potentials measured in the brain and measured
in vitro can be very different. Exposing a carbon paste electrode to
rats brain makes it more active. In Table 4 we give results for the

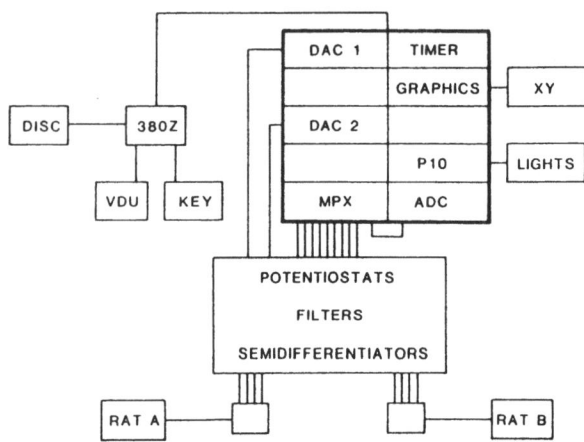

Fig. 18.   A schematic representation of the microprocessor-based
           equipment. DISC = Double two-sided 8" floppy disc drive.
           380Z = 380Z microprocessor unit. VDU = Visual display
           unit. KEY = keyboard. XY = X-Y plotter. The interface
           (enclosed by the thicker lines) contained a TIMER, two
           digital to analogue converts (DAC 1 and DAC 2), an eight
           channel multiplexer (MPX), an analogue to digital converter
           (ADC), two input/output-port sockets (P10s) and some
           graphics boards (GRAPHICS).

oxidation potential for a number of brain metabolites for untreated
electrodes, for electrodes exposed to rats brain (brain treated) and
for electrodes which had been cycled at 50 Hz between 0 and 3 V for
several seconds (electrically pre-treated). It can be seen that very
misleading identifications would be made if one relied on the potentials observed in in vitro studies for untreated electrodes.

     Identification of the compounds can be assisted by injecting
known compounds down a cannula round the implanted electrode. Using
this technique we have shown that peak 1 of Figure 19 is mainly
caused by ascorbic acid and not by the catecholamines [24]. Injection
of the former always caused an increase in the height of the peak but
no change in shape or peak potential. Injection of different
catechols produced an increase in current in this potential region
but the shape of the peak was altered and the peak potential shifted
to higher voltages. Similarly injection of 5-hydroxyindole acetic
acid (5HIAA) caused an increase in peak 2 with no change in shape
whereas injection of 5-hydroxytryptamine or glutathione shifted the
peak potential some 30 mV more positive. Hence it seems likely that
5HIAA is responsible for peak 2.

     Our conclusion that peak 1 is not caused by catechols is further
reinforced by injecting drugs into the rats which are known to

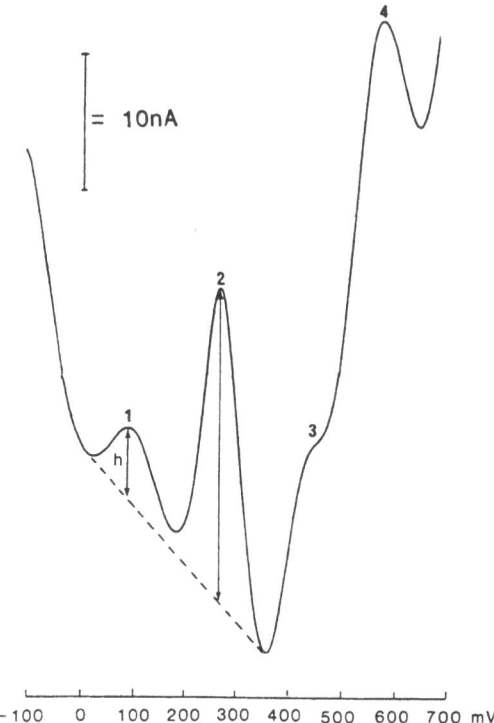

Fig. 19. A typical semi-differentiated voltammogram obtained with a
          carbon paste electrode in the striatum of an unanaesthe-
          tized rat. h = peak height.

release neurotransmitters. For instance injection of haloperidol
does not change the height of peak 1, although more current was
observed between peaks 1 and 2. This finding accords with the conclusions of Gonon et al.[22], who used carbon fiber electrodes which
could resolve the ascorbate peak from the catechol peak.

Furthermore we found that on injecting the rats with 3,4-di-
hydroxyphenylalanine (DOPA), again there was an increase in current
between peaks 1 and 2. Using the microprocessor one can easily
construct a difference voltammogram between the currents before and
after the injection. This is shown in Figure 20. The larger peak is
the oxidation of unchanged DOPA; the smaller peak is probably caused
by a methoxy derivative of the DOPA metabolism.

Because peak 1 is mainly caused by ascorbic acid, to monitor the
changes in concentration of the more interesting catechols one has to
construct difference voltammograms. A typical example from our
recent work is shown in Figure 21. The difference voltammogram shows
the effect of injecting haloperidol. The first peak is centred at

Table 4. Oxidation Potentials at Carbon Paste Electrodes [a)

Compound	Untreated Electrode	Electrically Pretreated Electrode	Brain Treated Electrode
Ascorbic acid	307 ± 20	25 ± 15	98 ± 10
3,4-Dihydroxyphenyl alanine (DOPA)	280 ± 8	117 ± 14	165 ± 3
Dopamine (DA)	157 ± 10	138 ± 13	127 ± 2
3,4-Dihydroxyphenyl acetic acid (DOPAC)	300 ± 12	130 ± 10	138 ± 5
Noradrenaline (NA)	217 ± 10	140 ± 12	155 ± 3
5-Hydroxytryptamine (5HT)	308 ± 6	272 ± 8	277 ± 3
5-Hydroxyindole acetic acid (5HIAA)	323 ± 3	270 ± 12	278 ± 2
3-Methoxytyramine	440 ± 12	370 ± 10	375 ± 2

Notes
a) The oxidation potentials are reported in mV versus $Ag/AgCl/Cl^-$
    (3M) reference electrode; they were obtained using LSV with SD
    at a pH of 7.4 at 37°C and using a sweep rate of 10 mV s^{-1}.
    Three different electrodes were used.

150 mV which is the potential for oxidizing cathechols; in this case
it is probably dihydroxyphenyl acetic acid (DOPAC) [23]. The second
peak in the difference voltammogram is centered at 360 mV which
corresponds to the oxidation of methylated cathechols. The time
course of these two peaks is shown in Figure 22. The concentrations
rise to a maximum after two hours and decline back to base line over
a 24 hour period. We estimate that the background concentration of
DOPAC being measured is some 20 μmol dm^{-3} and that the injection of
haloperidol increases this concentration to about 50 μmol dm^{-3}.

The in vivo measurement and monitoring of the chemical activity
of the brain is one of the most challenging problems for modern
electroanalytical chemistry. However these measurements are vital to
increase our understanding of the chemical effects of drugs and the
linkage between brain chemistry and behavioural response. The experiments may be difficult, the interpretation even more so, but this is
an area which urgently demands the closest collaboration between
electrochemists, neurophysiologists and psychologists.

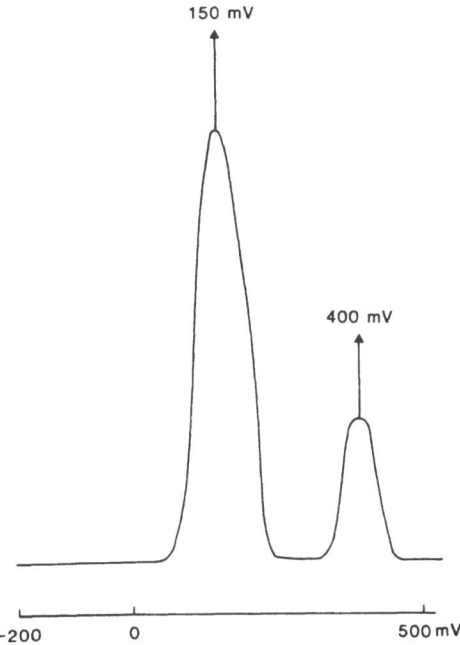

Fig. 20.   Difference voltammogram constructed by subtraction of the
           voltammogram before DOPA administration (50 mg/kg i.p.)
           from that recorded 90 min after injection.

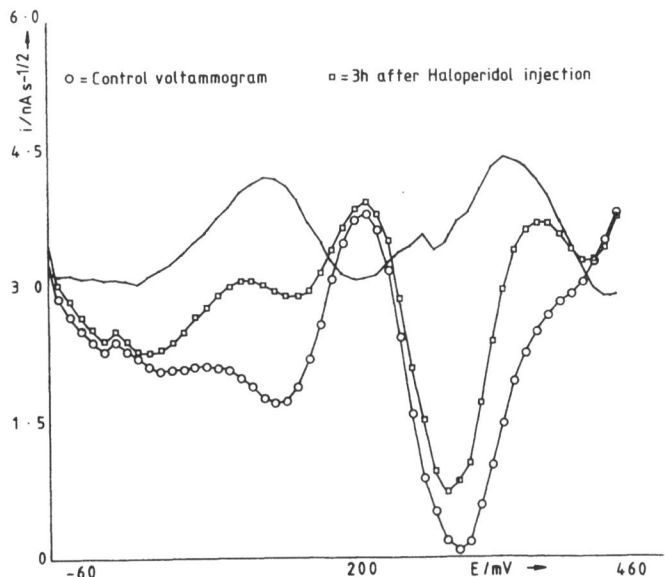

Fig. 21.   The effect of haloperidol (0.5 mg/kg i.p.) on part of the
           voltammogram recorded in the striatum of an unanaesthetized
           rat.   The centred zero indicates difference current = 0.

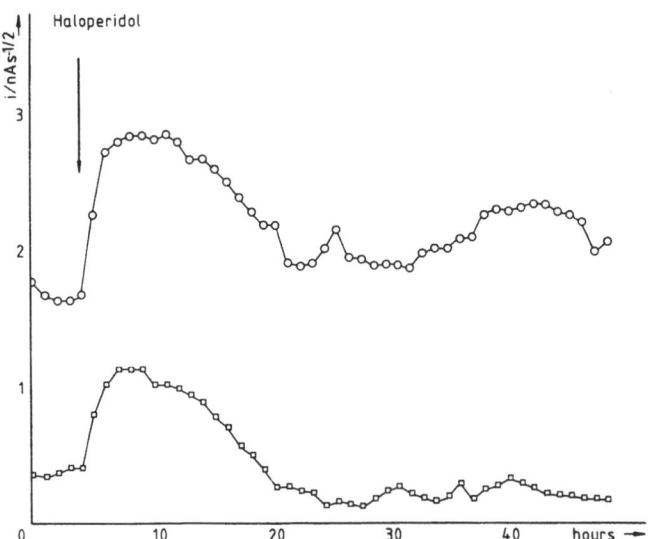

Fig. 22.   The time courses of the changes in the current, recorded in
           the striatum, at the oxidation potentials of the catechols
           (o) and the methylated catechols (□) after the injection of
           haloperidol (0.5 mg/kg i.p.) into unanaesthetized rats.
           Each value is the average of five measurements taken at 12
           min intervals.

Acknowledgements

        First we thank the SERC, MRC, ARC, BTG and Pharmacia for
financial support. Secondly we acknowledge the contributions made by
various research workers where results are reported in this article:

Diffusion Layer Titration of Proteins ... ... Dr P. Wood
                                                   Mr R. Svanberg
Membrane Electrodes   ... ... Dr W. N. Brooks
                                                   Dr P. Barron
                                                   Dr E. Hall
In Vivo Studies of Brain Chemistry   ... ... Dr W. Beck
                                                   Dr R. O'Neill
                                                   Mr R. Grunwald
                                                   Dr N. J. Goddard

REFERENCES

1. W. J. Albery and M. L. Hitchman, Ring-disc Electrodes, Clarendon
    Press, Oxford, p.99 (1971).
2. J. Yamada and H. Matsude, J.Electroanal.Chem., 44:189 (1973).

3. B. Fleet and C. J. Little, J.Chromatogr.Sci., 12:747 (1974).
4. L. C. Clark, Trans.Amer.Soc.Art.Int.Org., 2:41 (1956).
5. W. J. Albery, W. N. Brooks, S. P. Gibson, and C. E. W. Hahn, J.Appl.Physiol., 45:637 (1978).
6. W. J. Albery, W. N. Brooks, S. P. Gibson, M. W. Heslop, and C. E. W. Hahn, Electrochimica Acta, 24:107 (1979).
7. W. J. Albery, C. E. W. Hahn, and W. N. Brooks, Br.J.Anaesth., 53:447 (1981).
8. W. J. Albery and P. Barron, J.Electroanal.Chem., 138:79 (1982).
9. W. J. Albery and C. M. A. Brett, J.Electroanal.Chem., submitted for publication.
10. W. N. Brooks, C. E. W. Hahn, P. Foëx, P. Maynard, and W. J. Albery, Br.J.Anaesthesia., 52:715 (1980).
11. J. W. Severinghaus and A. F. Bradley, J.Appl.Physiol., 13:515 (1958).
12. P. R. Gifford and S. Bruckenstein, Anal.Chem., 52:1024 (1980).
13. I. Bergmann, Nature, 218:266 (1968).
14. P. T. Kissinger, J. B. Hart, and R. N. Adams, Brain Res., 55:209 (1973).
15. J. C. Conti, E. Strope, R. N. Adams, and C. A. Marsden, Life Sci., 23:2705 (1978).
16. R. F. Lane, A. T. Hubbard, K. Fukanagu, and R. J. Blanchard, Brain Res., 114:346 (1976).
17. R. F. Lane, A. T. Hubbard, and C. D. Blake, J.Electroanal.Chem., 95:117 (1979).
18. R. F. Lane and A. T. Hubbard, Anal.Chem., 48:1287 (1976).
19. W. J. Albery, T. W. Beck, W. N. Brooks, and M. Fillanz, J.Electroanal.Chem., 125:205 (1981).
20. M. Goto and D. Ishi, J.Electroanal.Chem., 61:361-365 (1975).
21. P. Dalrymple-Alford, M. Goto, K. B. Oldham, Anal.Chem., 49:1390-4 (1977).
22. F. Gonon, M. Burke, R. Cespuglio, M. Jouvet, and J. F. Pujol, Brain Res., 223:69 (1981).
23. R. D. O'Neill, M. Fillenz, W. J. Albery, and N. J. Goddard, Neuroscience, submitted for publication.
24. R. D. O'Neill, R. A. Grunwald, M. Fillenz, and W. J. Albery, Neuroscience, to be submitted.

VOLTAMMETRIC DETECTORS FOR HPLC AND OTHER ANALYTICAL

FLOW-THROUGH SYSTEMS

Antonín Trojánek

J. Heyrovský Institute of Physical Chemistry and
Electrochemistry, Czechoslovak Academy of Sciences,
118 40 Praha 1

The following paper gives a brief account of the properties, design and construction of voltammetric flow-through detectors for liquid chromatography (HPLC) and continuous flow analysis.

Voltammetric detectors represent the most important group of electrochemical detectors. Their popularity among electro-chemical detectors is so great, that the term "electrochemical" has become almost synonymous with "voltammetric" detection.

IMPORTANT PROPERTIES

Before starting to discuss concrete examples of detector construction, it is instructive to examine some of the properties which are important for their application. The most important in this respect are their:

a)   selectivity
b)   sensitivity
c)   dynamic range
d)   stability, reproducibility
e)   dynamic behavior

a) selectivity is usually defined as a dependence of the detectors' sensitivity on the type of detected compound. There are universal detectors (responding to a wide range of compounds to about the same extent) and specific ones (that respond much more sensitively to certain compounds than to others). Voltammetric detectors belong among specific ones and one of their greatest merits is, that their selectivity can easily be tuned by adjusting the value of the

23

working potential. Although constant-potential amperometry is the
only technique which has been widely used in practice, it is advis-
able in some cases for the sake of higher selectivity to apply voltammetric techniques, yielding response in the form of current peaks
(e.g., differential pulse voltammetry, AC voltammetry, square-wave
etc.). Those techniques make it possible to choose that value of
working potential to ensure the maximum sensitivity of the detector
towards the selected component in the mixture. As a practical example
the selective differential pulse polarographic LC detection of imper-
fectly separated nitrocompounds can be shown[1]. In Figure 1 is
depicted the response of the universal UV detector during analysis of
o-nitrophenol (ONPh) mixture. Chromatographic separation of the last
couple was, under applied experimental conditions, very poor and
corresponding compounds formed a mixed peak of no analytical value.
Fortunately these compounds differ electrochemically enough in their
half-wave potentials, so that it was possible to detect them selectively by means of differential pulse polarography at suitable working
potentials (Figure 2). Analogous advantages (together with enhanced
sensitivity) are offered by the application of phase-selective
sampled alternating current measurement[2]. For application in HPLC
a modification of the square-wave technique has been developed in
which the potential of the working electrode is periodically scanned
through the required potential range[3]. Substrate selectivity is
remarkably increased in this way, since the components of the original mixture are spread more widely along the potential axis.

     b) Sensitivity, in general defined as the slope of the response
signal-concentration dependence, is in the case of the flow-through
voltammetric detectors a function of the working potential, flow rate
and the area of the working electrode. The role of the working
electrode potential is self-evident and need not be discussed here.
Since the value of the measured current signal in general increases
with the flow rate and the area of the working electrode, it seems to
be logical to achieve increased sensitivity by increasing these two
parameters. Possibilities of increasing the flow rate in the whole
flow-through system are rather limited, since in LC the choice of
flow rate is subordinated to the chromatographic process and in
continuous flow analysis an increase of the flow rate would mean an
excessive consumption of reagents. More successful in this respect
is the application of "turbulent flow" detectors, where very effective mass transport towards the electrode is attainable. Of these,
the "wall-jet" and detectors containing turbulent tubular electrodes
are worth mentioning and will be discussed later on.

     When considering the total effect of increasing the area of
electrode, it should be kept in mind that increasing the magnitude of
the sensor's signal itself is of no practical value; it can be easily
achieved by electronic amplification. The fundamental quantity in
practically useful sensitivity is the value of the signal-to-noise
ratio, where as noise all the changes of the output signal which do

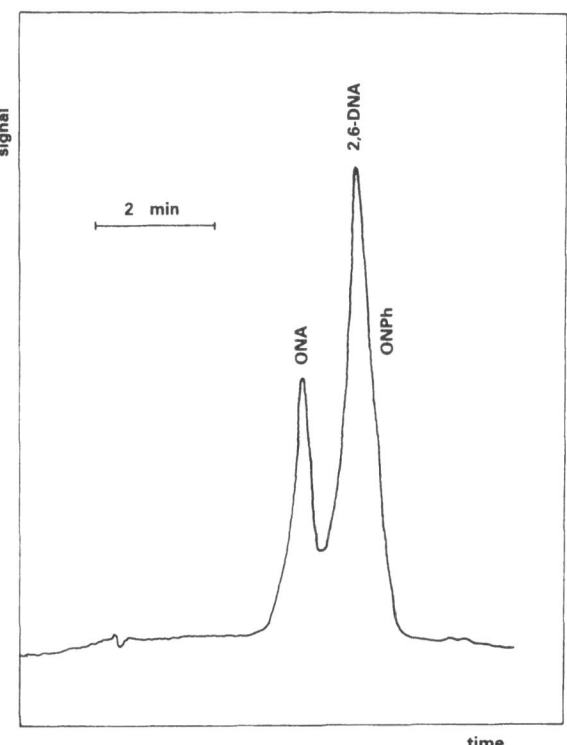

Fig. 1. Chromatographic separation of o-nitroaniline,
         2,6-dinitroaniline, o-nitrophenol mixture. 25 cm Separon
         C18 column. Mobile phase: acetate buffer pH 4.5, 50%
         methanol, 1 ml/min. Response of the universal UV detector.

not carry information about the input function (concentration) are
considered. Since the amplitude of the noise is usually directly
proportional to the area of the working electrode and the analytical
signal just to the square root of the area, increasing the area of
the electrode leads, in most cases, to a decrease in signal to noise
ratio (and, in consequence, to an increase of detection limit). The
advantage (in view of the S/N ratio) of employing small electrodes
was verified for both voltammetric and polarographic detectors[4].
At polarographic detectors the best S/N ratio was obtained with small
mercury drops at a fast drop rate (facilitated by the use of a horizontally placed capillary).

     Signal to noise ratio can be, on the other hand, remarkably
improved by application of pulse and AC techniques.

     c) Dynamic range is another important property governing the
practical applicability of the detector. For voltammetric detectors

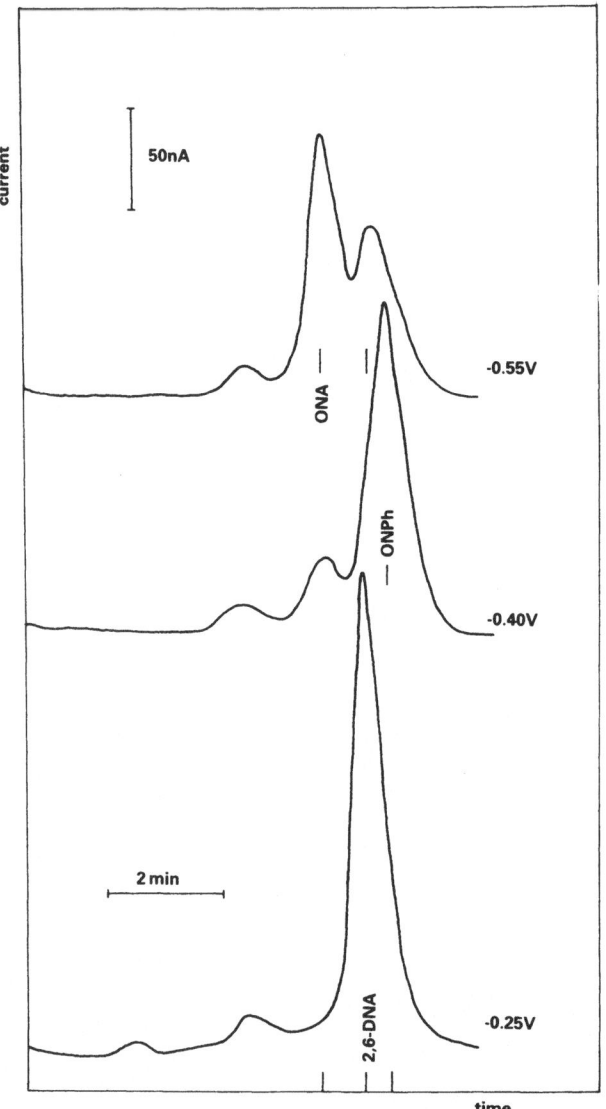

Fig. 2. Response of the polarographic detector. Differential pulse
          polarographic technique, pulse amplitude 25 mV, constant
          potential values denoted. Other conditions the same as in
          Figure 1.

linear dynamic range of about five orders of magnitude is usually
given. This compares favorably with UV ($\sim 5.10^3$), refraction ($\sim 10^3$)
and fluorescence detectors ($\sim 10^3$-10^6)[5].

     d) Stability and reproducibility of the response is a weak point
of voltammetric detectors in general. It is connected primarily with

the activity of the sensor (electrode) which is directly exposed to
the flowing medium and consequently often suffers from passivation
and gradual loss of its activity. Numerous methods to prevent the
decrease of sensor activity have been worked out (mechanical,
chemical, electrochemical), but none of them is quite reliable.
Detectors employing dropping mercury electrodes are, of course, free
from these problems.

     e) Dynamic behavior is, in contrast to the properties discussed
above, a complex property of the whole detection system and not of
the sensor only. Problems connected with peak broadening and response
rate, reflecting dynamic properties, were thoroughly studied by
several workers[6-8]. One of the most important conclusions is that
in order to construct detectors with a fast response it is necessary
to design a cell with small internal volume. But it should be borne
in mind that, besides geometrical volume, other parameters play
important roles also; namely flow pattern in the cell, shape of the
cell, sensor etc. Data presented in the literature regarding the
response properties are not uniform in their definitions and concepts
such as response, rinse, dead, hold-up and wash-out volumes. In this
respect one should take the values of declared volumes provided by
various manufacturers with a pinch of salt since they usually correspond to the geometrical volume and do not reflect the really dynamic
behavior of the detector. For instance, practical measurement with
the commercial polarographic detector PAR-310 with declared geometrical volume less than 1 μl revealed dynamic behavior corresponding to
about 80 μl wash-out volume[9].

     The most important dynamic parameter is the time constant of the
detection system. Since it is, in general, flow rate dependent, the
calculated response volume seems to be the most suitable parameter
for characterization of the dynamic properties and for comparison of
various designs[7].

     In addition to the requirements discussed above, there are some
special considerations in voltammetric detector construction, arising
from the electrochemical point of view. The need to ensure minimum
time constant of the electrode system (especially for application of
pulse techniques) and the need to achieve homogeneous distribution of
the potential require that the electrodes be placed symmetrically
close to each other. The proper functioning of the poteniostatic
equipment requires the positioning of the reference electrode as
close as possible to the working electrode. Meeting these requirements (which are in contradiction with the previously stated ones for
the minimum internal volume) creates certain construction problems.
Thus it is not surprising that there are numerous constructions in
use in some of which these requirements are ignored. Reference (in
some cases together with auxiliary) electrodes are often placed in
the waste vessel, which is electrolytically connected with the working electrode. The application of a three - electrode configuration

is, in such cases, rather formal and enables only the utilization of reference electrodes of a high internal resistance. Calibration curves are then usually non-linear as a result of the potential changing in regions of higher concentration (higher current values).

DETECTORS

Detectors for flow-through voltammetric measurements are currently very numerous and their construction strongly depends on the material chosen for preparation of the working electrode. Whereas the shape of mercury electrodes in polarographic detectors is given by the shape of the mercury meniscus, the choice of shapes of solid electrodes is practically unlimited. Electrodes in the forms of discs, foils, rods and grids are commonly used, while typical flow-through forms are porous and tubular electrodes. For HPLC, where fast response and ease of exchange of passivated electrodes is very important, thin-layer detectors are usually preferred.

Mercury Electrodes

In spite of several drawbacks which arise from the application of mercury as an electrode material (limited anodic range, chemical reactivity and toxicity, difficult handling etc.), mercury electrodes are widely used. The reason, apart from their extensive cathodic range of polarization is that they enable the utilization of a huge amount of experimental material, collected since polarography was invented.

Dropping Mercury Electrode (DME)

In addition to the merits given above, the DME provides a constantly renewed electrode surface, essentially eliminating the problem of electrode surface contamination. Periodical dropping of mercury results, on the other hand, in current oscillations, which contribute to the level of noise and complicate the evaluation of polarograms and possible further handling of the signal. Suppression of the oscillations necessitates some form of damping. The use of RC filters leads to distortion of the recorded response curves and reduces substantially the response rate. The most generally useful method of eliminating oscillations involves current sampling at the end of a drop life, determined usually by the action of a mechanical knocker or by the frequency of auxiliary potential pulses (of about 1 volt amplitude), applied to the electrode[10]. The latter principle is based on the potential dependence of the surface tension. If a pulse of proper polarity at the end of the natural drop life is applied, surface tension sharply decreases and the drop falls.

Another approach is the method developed in our laboratory, based on the application of a peak detector, which is periodically reset and made ready to accept a new maximum value. This method does not require synchronization of drop disconnection with the start of current sampling[11].

The earliest studies with flow-through polarographic detectors were done in the fifties in conjunction with so-called chromatopolarography, that is with liquid column chromatography, utilizing polarography as a detection method[12]. Another area where flow-through detectors were introduced first, was the construction of automated analyzers (eg. for determination of oxygen in lake water[13], activated sludge, etc.[14]). During the following years a great number of different flow-through detectors was constructed, which were gradually becoming more and more distinct from detectors applied in quiescent solutions. The polarographic flow-through cell of Blaedel and Strohl (from the sixties) is a good example from the beginning of this development[15]. Its cylindrical body was made of Teflon, and an axial hole was bored in it through which analysed solution flowed. DME and reference electrodes were placed perpendiculary to the direction of the flow. The detector exhibited, due to the relatively small internal volume, rather fast response-according to the given data; current values of 90% and 99% of steady state being obtained on passing 500 and 750 µl respectively of sample solution. Construction is simple, but requires exact machining of the material.

Another, more recently developed design is shown in the Figure 3[16]. The detector body consists of Perspex. The detection occurs at DME in a cylindrical space with a volume of about 1 µl. The detector exhibits good geometry of the electrodes, while into the vertical side channel a third electrode can be inserted. Standard deviation (which is for ideal mixer conditions equal to the response volume) was found to be 10-30 µl.

Another model can be seen in Figure 4[17]. The direction of the flow of the analysed liquid is again perpendicular to the direction of the mercury flow. Such arrangement ensures the most efficient transport and consequently the highest sensitivity. The analysed solution is split into two branches, ensuring the electrolytic connection of the working electrode with reference and counter electrode. Conical grinding of the DME enables the achievement of low effective working volume of about 2-5 µl.

Another detector (Figure 5), designed for Technicon continuous flow analyzers, is simple in construction (with reference electrode placed outside the cell, in waste bottle)[18]. The simplicity of the cell was obtained at the cost of its dynamic properties. Nevertheless, the response times obtained with this detector were found to be sufficient for the given purpose.

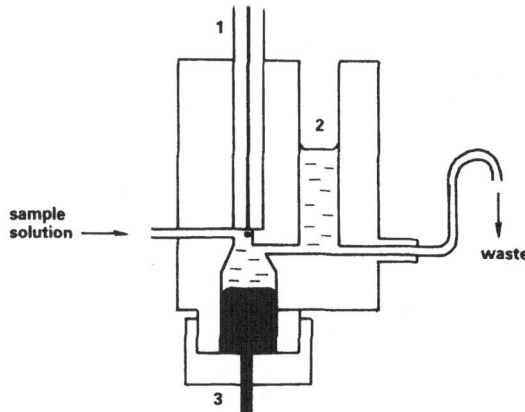

Fig. 3.   Polarographic detector.   1 – DME, 2 – channel for placement
of the third electrode, 3 – pool reference electrode.

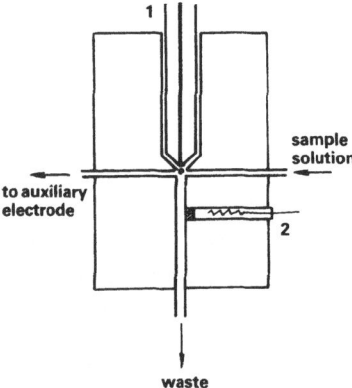

Fig. 4.   Polarographic detector.   1 – conically ground capillary of
the DME, 2 – reference electrode.

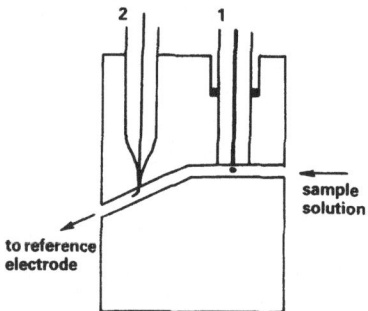

Fig. 5.   Polarographic detector.   1 – DME, 2 – auxiliary platinum
electrode.

The detector for amperometry, in a two electrode configuration, of Feher and Pungor (Figure 6) is also very simple. The electrodes can be placed wherever is considered convenient.

The detector of Wasa and Musha in Figure 7 has a body of sandwich construction, which makes it possible to achieve complicated internal cavities[20].

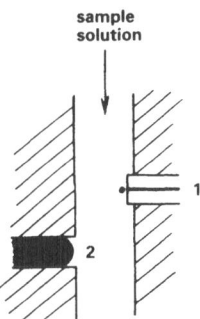

Fig. 6. Polarographic detector. 1 – DME, 2 – meniscus of mercury reference electrode.

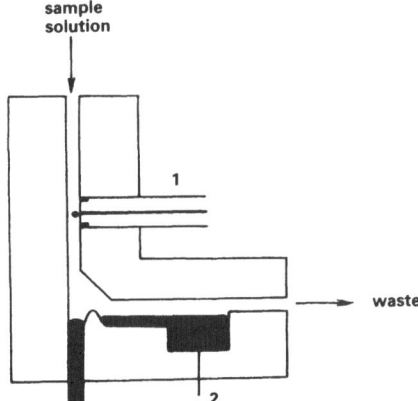

Fig. 7. Polarographic detector. 1 – DME, 2 – pool reference electrode.

An excellent performance is exhibited by the detector of Michel and Zátka; introduced recently[21]. Detector incorporates a horizontal dropping mercury electrode and a solid movable piston, situated against the capillary orifice. The presence of a solid body against the orifice hinders the free expansion of the drops and makes it possible to achieve short drop times, which can be adjusted within some limits by adjusting the orifice-piston distance by the fine screw. For instance, when the distance is 70 µm, the drop time is about 50 ms (and the internal geometrical volume is about 0.5 µl).

Greatest advantage of this design is the low internal volume. The
relatively high frequency of data acquisition is a good premise for
the exact recording of the signals and for the suppression of undesirable oscillations of the current (for suppression the time constant of the common types of recorder is sufficient). The drawback
is the rather complicated construction and with the model described,
there have been observed disturbances in the recorded current at flow
rates above 1 ml/min. A detector of this type is commercially
available, manufactured by the French firm Solea-Tacussel[22].

The present trend towards the utilisation of detectors with a
very short drop time (which is favorable in respect of the signal/
noise ratio) is exemplified in the detectors designed in Holland by
Hanekamp and Bos[23,24]. Their detectors utilize in principle,
horizontally mounted capillaries, and short drop times are achieved
by means of a movable pin, or, more recently, by suitable positioning
of the electrodes. Conical grinding of the capillary and of the
glass tip of the reference electrode enables the internal volume of
the detector to be minimised. With this type, response volumes of
8-15 μl were obtained depending on the flow rate.

From the survey of polarographic detectors, given above it is
evident, that the application of the DME enables the construction of
detectors having very good properties. But their construction
requires exact machining of the material, often including grinding of
the capillary. In this respect the commercial detector manufactured
by the American firm PAR (PAR-310) seems to be very elegant[25]. A
three - electrode system is placed in the vessel filled with a suitable base electrolyte; the size and shape of the electrodes is practically identical with those, used for measurement in quiescent
solutions (the glass capillary of the DME is ground conically). The
analyzed solution is delivered through the base electrolyte onto the
dropping electrode. An interesting point is that the given principle
has been already described in the year 1970[26].

The detector developed in our laboratory is very simple in
construction and the principle of operation is based on the introduction of analyzed solution into a chamber filled with mercury[27].

Static Mercury Electrodes

Detectors with static mercury electrodes resemble in construction detectors with solid electrodes. Due to the loss of the periodically renewed surface their application is advantageous in excep-
tional cases only (for example, when it is necessary to work at
higher cathodic potential). Since the stability of the hanging drop
in the flow is rather limited, a mercury pool electrode is often
preferred[28]. There are also some constructions where the stability
of the mercury meniscus was improved by its coverage by a suitable

membrane[29]. Coverage really does improve the stability, but at the
cost of response rate and reproducibility. New horizons in the
application of static electrodes opened with the introduction of the
PAR-303 static mercury electrode, which gives the possibility of
constructing detectors in which the electrode surface can be renewed
at any selected moment.

Solid Electrodes

As has already been mentioned, the typical forms of solid flow-
through electrodes are porous and tubular electrodes.

Porous Electrodes

In detectors equipped with porous electrodes analysed solution
is percolated through the electrode material. Due to the large
surface area of the working electrode, high current values are
obtained even at low concentrations of electroactive species.
Unfortunately, this large analytical signal is, for the same reason,
accompanied by high background signal. In accordance with our pre-
vious discussion, the application of a large surface electrodes is
generally unfavorable because of the S/N ratio. Together with the
impossibility of cleaning the electrode surface this is the reason
why is the application of porous electrodes in practice so rare.

In the past, porous electrodes were usually made of sintered
silver or were prepared in the form of columns packed with suitable
material. Recently (about 2-3 years ago) the interest in the application of porous electrodes increased as a new porous material became
available. The material is called reticulated vitreous carbon (RVC)
and it is a porous glassy carbon, available in various forms. Flow-
through detectors of a simple design can be constructed in the form
of a column filled with RVC, with other two electrodes immersed in
the waste bottle[30]. Construction is very simple, but again at the
cost of potential homogeneity, with a potential gradient along the
column. Detectors of better design (and of more complicated construction) can be seen in the Figure 8[31].

Designs based on various numbers of RVC disks can be made to
work (according to the required degree of conversion). With this
detector the signal-to-background ratio was greatly improved by
differential current measurements in a stopped-flow procedure.
Measurement of the current difference in the flow and in the stopped
flow discriminates against the nonconvective components, which has
been reported as the major contribution to the steady-state background currents. However, the long cycling period (about a minute)
of the stopped-flow procedure precludes its use in chromatography and
analyzers. For such applications, pulsed flow procedures have been

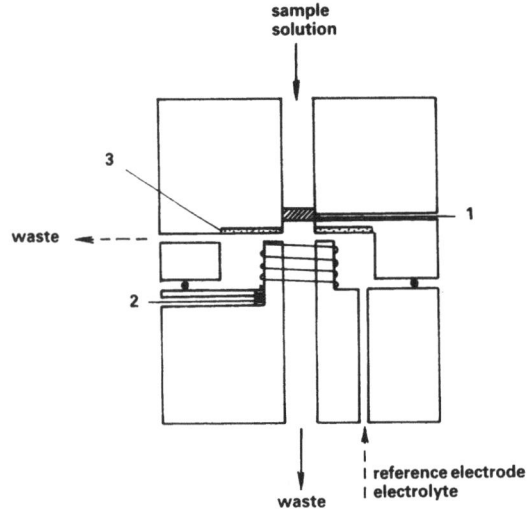

Fig. 8.   Detector with RVC electrode. 1 - lead to the RVC working
          electrode, 2 - lead to the reference electrode, 3 - ion
          exchange membrane.

worked out, where the alternating signal (originating from the pul-
sation of the flow) is electronically processed and isolated from the
non-alternating signal of the background.

Tubular Electrodes

        Tubular electrodes are generally formed by a tube made of con-
ductive material. Platinum and graphite, are commonly used though
the application of glassy carbon is rare, due to its difficult
machining. The inner surface of the tube forms the surface of the
working electrode.

        In spite of simplicity and excellent hydrodynamic properties,
tubular electrodes are not commonly used due to the difficulties
connected with polishing and cleaning of their inner surface in
detectors of small hold-up volumes.

        In the Figure 9 is presented a sophisticated model with a
streaming liquid junction between electrodes[32]. Such an arrangement prevents contamination and choking of the solid junctions
usually employed.

        In the Figure 10 a detector with a so-called "turbulent tubular
electrode" is presented[33]. The cell is equipped with a strip
Teflon stirrer, whose axis of rotation is parallel to the axis of the

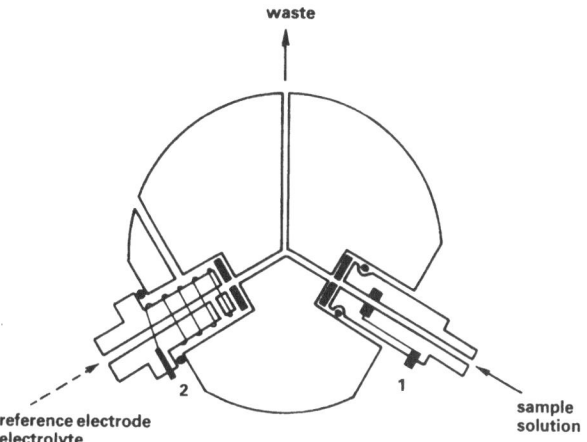

Fig. 9. Tubular electrode detector with flowing junction between
         electrodes. 1 - lead to the tubular platinum electrode,
         2 - lead to the reference electrode.

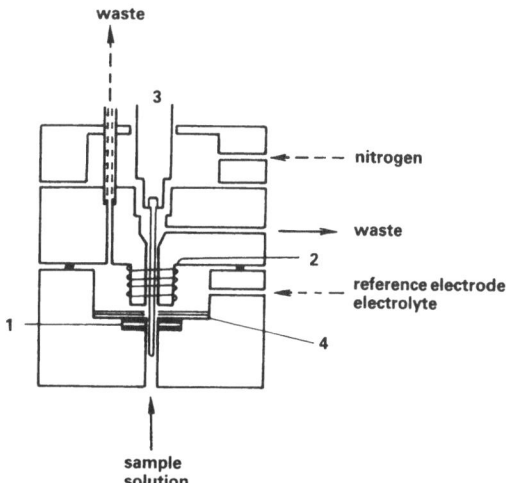

Fig. 10. Detector with turbulent tubular electrode. 1 - lead to
          the tubular platinum electrode, 2 - lead to the reference
          electrode, 3 - stirrer holder, 4 - ion exchange membrane.

tubular electrode. With the aid of the mechanical stirrer high mass
transport rates are attainable due to a turbulent flow regime. The
detector exhibits the basic properties of detectors equipped with
rotating electrodes.

     The American firm Chromatix manufactures for chromatographic
purposes a detector CMX-20 with tubular carbon/polymer-matrix working

electrode. According to the data available from manufacturer[34], the working electrodes have an expected lifetime of several months (after that they can be easily replaced). Electrodes come ready to use and do not require prior preparation or polishing - in contrast to glassy carbon electrodes.

Thin-Layer Cells

In LC, where the requirement of minimum internal volume of a detector is of prime importance the most popular are thin-layer detectors. Thin-layer detectors have been constructed with the stream directed perpendicular to the electrode surface or with the stream flowing parallel to the electrode embedded in the channel wall.

Detectors of the first type are called 'wall-jet' detectors and this name is quite realistic, as can be seen in the Figure 11 where the well-known detector of Fleet and Little [17] is presented. (An analogous model has been developed by Pungor and co-workers in the sam year 1974)[35]. The solution is introduced via a nozzle and impinges normally onto the planar disk electrode. The stream is then split into two opposed channels, connecting the working electrode with two other electrodes. The distance between the nozzle tip and the electrode surface is adjustable and determines the effective cell volume. The typical cell volume (geometrical) is less than 1 µl.

A similar model is commercially available (the Model LCA 15 manufactured by the British firm EDT Research) with the working electrode made of glassy carbon and the body of KE1-F fluorocarbon. The internal geometrical volume of the detector is 0.5 µl.

Wall-jet detectors have several useful characteristics:

- high sensitivity (due to the very efficient mass transport),
- freedom from surface adsorption (due to the mechanical washing effect of the impinging liquid).

In most cases the variable cell volume is of little practical use, since the minimum possible volume is usually required. The performance principles of detectors, where electrolysis takes place in the thin-layer of electrolyte, flowing parallel to the electrode surface, can be demonstrated on the commercially available thin-layer detector from Bioanalytical Systems (USA), schematically depicted in the Figure 12[36]. The detector works with a carbon paste or glassy carbon working electrode. The thickness of the layer is determined by the thickness of a Teflon film spacer (~50 µ) between two blocks of plastic. Since there is not enough space for location of the other two electrodes, they have to be placed outside of the thin-

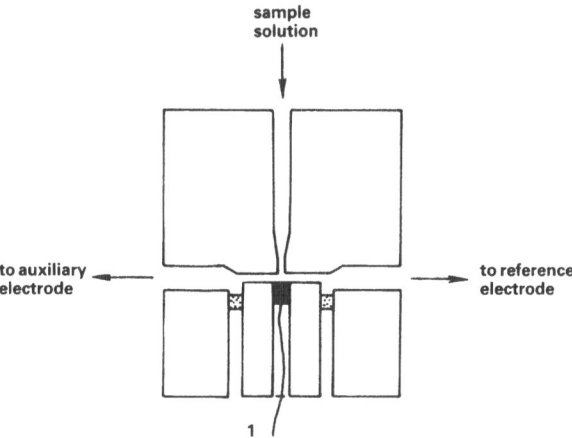

Fig. 11. Wall-jet detector. 1 - lead to the glassy carbon electrode.

layer region[37]. According to the manufacturer, this offence
against electrochemical principles has no negative effects on the
performance of the detector (due to the nanoampere level of the
measured currents). This detector is one of the most commercially
successful models on the market. Thin-layer detectors of this type
exist in many variants. For instance a detector was described,
containing two (or more) carbon-paste working electrode embedded in
the same wall[38]. Electrodes are maintained at different potentials
and the resulting current outputs are recorded separately. For
compounds that overlap chromatographically, but have different electrolysis potentials, the cell provides selective detection. A detector with two electrodes embedded in the opposite sides of the channel
enables an increase in detection sensitivity of reversibly reacting
species[39].

A flow-through detector of similar electrode geometry was
designed for capillary LC also. Due to the extremely low flow rates
(0.5-5 µl/min) and extremely small sample volumes (tenths of µl) in
this technique, there are extreme constraints on the cell volume. In
the above detector (which is, by the way, the only electrochemical
detector utilized in connection with capillary LC) was the cell
volume 0.15 µl[40].

In the Figure 13 the construction of a differential amperometric
detector is shown[41]. The detector consists of two identical thin-
layer cells, which are in contact with a compartment containing the
reference and auxiliary electrodes. A mobile phase is pumped through
the reference cell so that the recorded differential signal is free
of background noise, caused by the presence of electroactive impurities in the solvent.

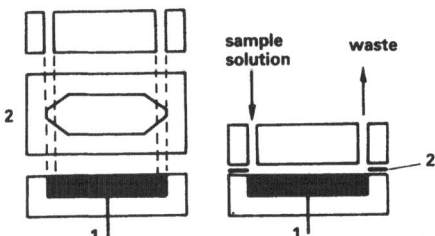

Fig. 12.   Thin-layer detector. 1 - lead to the glassy carbon or
           carbon paste electrode, 2 - teflon spacer.

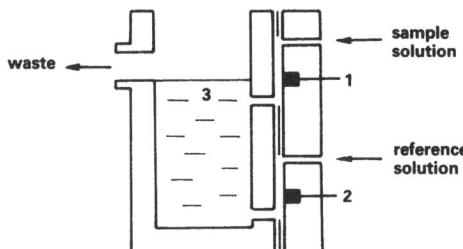

Fig. 13.   Differential thin-layer detector. 1,2 - leads to the
           glassy carbon electrodes, 3 - reference and auxiliary
           electrodes chamber.

     In the Figure 14 a cell is shown with a good symmetry of elect-
rodes. Due to the low resistance between electrodes this detector
could be used also for detection by the technique of differential
pulse voltammetry[42].

     The last type of amperometric flow-through detector to be
discussed are detectors with rotating working electrodes, oriented
perpendicularly to the flow.

Rotating Electrodes

     The introduction of moving mechanical parts in general compli-
cates detecting devices in respect of design and maintenance. On the
other hand, in some applications it brings certain merits. First, it
enables operation without sacrificing sensitivity at low flow rates.
As has been already mentioned, this feature can be of some interest
when a detector is applied in continuous flow analyzers, where
decrease in flow rate means decrease in reagent consumption. In
addition the measured current signal is (within usual range of flow
rates and speeds of rotation) practically independent of the flow
rate, since using a RDE, the thickness of the diffusion layer is
mainly determined by the rotation speed of the electrode[43, 44]. An
example of the construction is in the Figure 15. The rotating disk

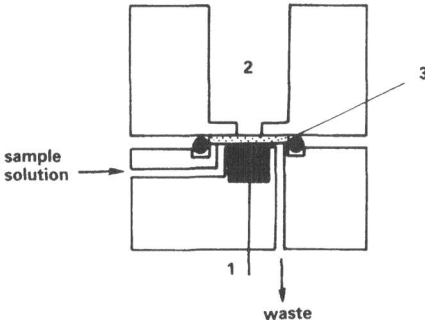

Fig. 14. Thin-layer detector. 1 - lead to the glassy carbon
          electrode, 2 - reference and auxiliary electrodes
          chamber, 3 - ion exchange membrane.

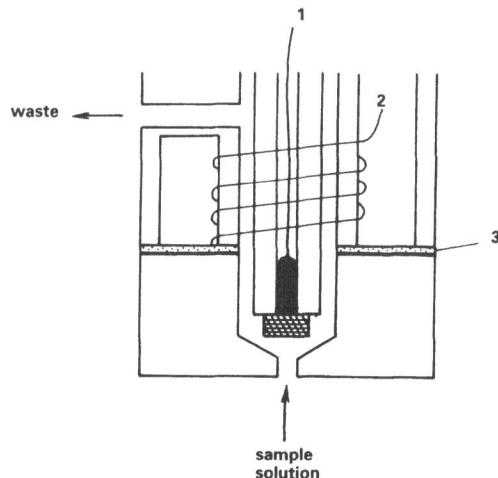

Fig. 15. Detector with rotating porous electrode. 1 - lead to the
          RVC disc electrode, 2 - lead to the reference electrode,
          3 - ion exchange membrane.

is made of RVC so that the efficient mass transport from the rotating
of the electrode is coupled here with the large surface area of the
porous electrode[45]. The high analytical currents are again
accompanied by a high background current. A suitable method for its
compensation seems to be "pulsed rotation technique". In this technique the rotation speed is periodically changed (pulsed) between two
values and the difference between corresponding current values is
plotted. In comparison with the previously mentioned method of
pulsed flow this approach has the advantage of not disturbing the
flow pattern in the liquid manifold until the analysed solution
reaches the sensor. The technique is, of course, again not applic-

able for systems where fast response is necessary (LC) due to the relatively long period of data acquisition (at least 20s).

CONCLUSION

The introduction of voltammetric detectors into flow-through analysis brings with it the possibility of recognising detected species according to their electrochemical properties. From the above brief survey it follows that properly designed voltammetric detectors exhibit excellent properties even if they are still far from the concept of the "black box detector" required by workers in practice. Future development can be expected in the simplification of mechanical design, the introduction of new voltammetric techniques and in the wider application of theoretical knowledge in detector design. Scientifically designed detectors, incorporating the results of theoretical studies are, in practice, still rare as are children born within the framework of family planning!

REFERENCES

1. A. Troják, unpublished results.
2. H. B. Hanekamp, W. H. Woogt, R. W. Frei and P. Bos, Anal.Chem., 53:1362 (1981).
3. R. Samuelson and J. Odea, J. Osteryoung, Anal.Chem.
4. H. B. Hanekamp and H. J. van Nieuwkerk, Anal.Chim.Acta, 121:13 (1980).
5. M. Varadi, Pure Appl.Chem., 51:1175 (1979).
6. H. Poppe, Anal.Chim.Acta, 114:59 (1980).
7. H. B. Hanekamp, P. Bos, U. A. Th. Brinkman and R. W. Frei, Z.Anal.Chem., 297:404 (1979).
8. J. C. Sternbert in "Advances in Chromatography", Vol. 2, ed., J. Calvin Giddings and R.A. Keller, Marcell Dekker Inc., N.4. (1966) p.205.
9. H. B. Hanekamp, private communication.
10. H. B. Hanekamp, W. H. Voogt and P. Bos, Anal.Chim.Acta, 118:73 (1980).
11. A. Troják and I. Holub, Anal.Chim.Acta, 110, 161 (1979).
12. W. Kemula, Rocz.Chem., 26:281 (1952).
13. C. P. Tyler and J. H. Karchmer, Anal.Chem., 31:499 (1959).
14. G. J. Alkire, K. Koyama, K. J. Hahn and C. E. Michelson, Anal.Chem., 30:1912 (1958).
15. W. J. Blaedel and J. H. Strohl, Anal.Chem., 36:445 (1964).
16. J. G. Koen, J. F. K. Huber, H. Poppe and G. den Boef, J.Chromatogr.Sci., 8:192 (1970).
17. B. Fleet and C. J. Little, J. Chromatogr.Sci., 12:747 (1974).
18. W. Lund and L. N. Opheim, Anal.Chim.Acta, 79:35 (1975).
19. Zs. Fehér and E. Pungor, Anal.Chim.Acta, 71:425 (1974).
20. T. Wasa and S. Musha, Bull.Chem.Soc.Jap., 48:2176 (1975).

21. L. Michel and A. Zátka, Anal.Chim.Acta, 105:109 (1979).
22. Solea-Tacussel, DELC Electrochemical detection system for HPLC, (1981)
23. H. B. Hanekamp, P. Bos, U. A. Th. Brinkman and R. W. Frei, Z.Anal.Chem., 297:404 (1979).
24. H. B. Hanekamp, P. Bos and R. W. Frei, J.Chromatogr., 186:489 (1979).
25. Princeton Applied Research, Electrochemical Accessories (1978/79)
26. E. Scarano, M. G. Bonicelli and M. Forina, Anal.Chem., 42:1470 (1970).
27. A. Trojánek, preparing for publication.
28. P. W. Alexander and S. H. Qureshi, J.Electroanal.Chem., 71:235 (1976).
29. E. Pungor, G. Nagy and Zs. Fehér, J.Electroanal.Chem., 75:241 (1977).
30. A. N. Strohl and D. J. Curran, Anal.Chem., 51:353 (1979).
31. W. J. Blaedel and J. Wang, Anal.Chem., 51:799 (1979).
32. W. J. Blaedel and Z. Yim, Anal.Chem., 50:1722 (1978).
33. W. J. Blaedel and G. W. Schieffer, Anal.Chem., 46:1564 (1974).
34. Chromatix, CMX-20 amperometric HPLC detector.
35. M. Váradi, Zs. Fehér and E. Pungor, J.Chromatogr., 90:259 (1974).
36. Bioanalytical Systems, detector LC-16.
37. P. T. Kissinger, C. J. Refshange, R. Dreiling and R. N. Adams, Anal.Lett., 6:465 (1973).
38. R. E. Shoup and P. T. Kissinger, Chem.Instr., 7:171 (1976).
39. R. J. Fenn, S. Siggia and D. J. Curran, Anal.Chem., 50, 1067 (1978).
40. Y. Hirata, P. T. Lin, M. Novotný and R. M. Wightman, J. Chromatogr., 181:287 (1980).
41. K. Brunt and C. H. P. Bruins, J. Chromatogr., 161:310 (1978).
42. G. W. Schieffer, Anal.Chem., 52:1994 (1980).
43. K. Brunt, C. H. P. Bruins, D. A. Doornbos and B. Oosterhuis, Anal.Chim.Acta, 114:257 (1980).
44. B. Oosterhuis, K. Brunt, B. H. C. Westerink and D. A. Doornbos, Anal.Chem., 52:203 (1980).
45. W. J. Blaedel and J. Wang, Anal.Chem., 52:1697 (1980).

SOME APPLICATIONS OF ELECTROCHEMICAL OXIDATION AS A DETECTION

TECHNIQUE IN HIGH PERFORMANCE LIQUID CHROMATOGRAPHY

A. J. Samuel and T. J. N. Webber

Shell Research Limited
Sittingbourne Research Centre
Sittingbourne, Kent ME9 8AG

INTRODUCTION

High performance liquid chromatography (HPLC) has been shown to
be a powerful separation technique though its range of application is
limited by the methods available for detecting the eluates. UV
absorption detectors provide sensitivities down to a few nanograms
for compounds with strong UV chromophores. Bulk-property detectors
such as refractive index detectors provide a nearly universal response but with sensitivities in the microgram or high nanogram
range. Fluorescence detection can yield picogram sensitivities but,
unless derivatization techniques are used, it is restricted to those
compounds with strong fluorophores. The electrochemical detector [1]
is a selective detector, the sensitivity and range of application
depending on the electroactivity of the compounds of interest.
Whilst the application of electrochemical detection in HPLC is becoming established for certain classes of compounds, such as biogenic
amines [2] and plant phenolic materials [3], its use for compounds
that are less easily oxidized has not been widely examined.

The aim of this paper is to describe the principal considerations for setting-up an electrochemical detection system for HPLC,
including the determination of electroactivity of organic compounds
using a novel microscale voltammetric technique, the choice of mobile
phase and the selection of instrumental approaches for optimum performance. The application of electrochemical detection to several
classes of less easily oxidized compounds, including sterols, non-
ionic surfactants and organic acids, is illustrated.

EQUIPMENT

(a) Electrochemical flow-cells

The EDT Research (London, Great Britain) wall-jet cell model
LCO3, refer Figure 1, or in later experiments model LCA-13 was used
in the majority of this work. These flow-cells incorporate a glassy
carbon working electrode onto which the eluent is forced through a
narrow orifice. A reference electrode (Ag/AgCl) and a stainless
steel counter electrode are located in close proximity to the working
electrode.

The Bioanalytical Systems Inc. (from Anachem, Luton, Great
Britain) flow cell TL-4 was used to evaluate carbon paste working
electrodes. The TL-4 is a two compartment cell with the working
electrode monitoring the eluent stream from the column as it flows
through a thin-layer gap inside the cell. Counter and reference
electrodes are held in a second compartment connected to the working
electrode via a wide bore tube.

(b) Potentiostats

The Bioanalytical Systems Inc. potentiostat (LC-2A) was used in
most of the work. The original instrument was modified to decrease
its sensitivity (x10) so that it could be used a high working potentials and high standing currents. This was achieved by changing the
feedback resistor of the current-to-voltage converter from $10M\Omega$ to
$1M\Omega$.

In some work an unmodified Princeton Applied Research Corp. (EG
and G Ltd., Bracknell, Great Britain) polarographic analyser (174A)
was used in its DC mode of operation.

(c) Cyclic voltammetry equipment

The Bioanalytical Systems Inc. cyclic voltammetry instrument
(CV-1A) was used with a conventional three electrode cell consisting
of a planar carbon paste or glassy carbon working electrode, a saturated calomel reference electrode and a platinum wire counter electrode.

(d) Chemicals

All chemicals used were Analytical Reagent grade. The chromatographic solvents, where available, were HPLC grade (Rathburn
Chemicals, Walkerburn, Great Britain). Other solvents were of
reagent grade and were used without further treatment.

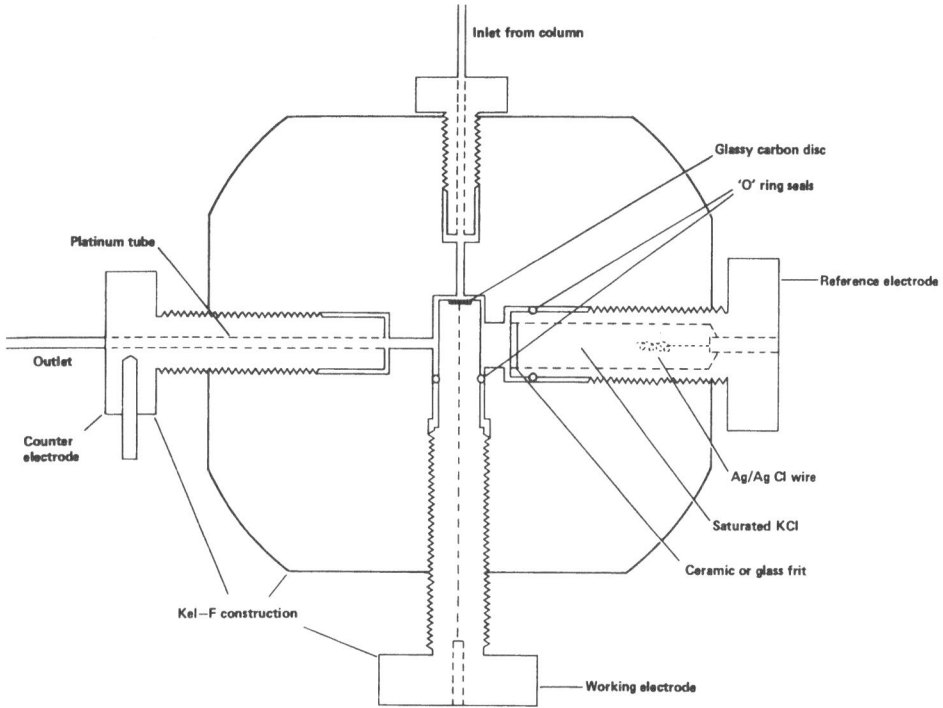

Fig. 1. EDT Research wall-jet electrode cell, LC03.

(e) Column-packing materials

The following packing materials were used:

Lichrosorb S1-60   (5 µm))
)BDH Ltd., Poole, Great
Lichrosorb RP-8 and RP-18 (10 µm))Britain

Hypersil ODS (5 µm)   Shandon Southern Products
                                   Ltd., Runcorn, Great Britain

Partisil-10 PAC   (10 µm) Whatman Labsales Ltd.,
                                   Maidstone, Great Britain

The HPLC columns were constructed from 0.10 m or 0.20 m x 4.5 mm ID Apollo liquid chromatography stainless steel tubing (Magnus Scientific, Sandbach, Great Britain) and were packed in the laboratory using a carbon tetrachloride slurry technique [4] at a packing pressure of 31.0 MPa (4500 psig).

(f) Liquid chromatographic system

    The liquid chromatograph was laboratory assembled from commer-
cially available components. The solvent delivery system consisted
of a Metering Pumps Ltd. (London, Great Britain) liquid pump, type
HM. Pulse damping was effected by the combination of a Bourdon type
pressure gauge and a coiled length (25 m x 0.25 mm ID) of stainless
steel capillary tubing. The electrochemical flow-cell was connected
directly to the HPLC column.

    A Cecil CE212 variable wavelength UV detector (Cecil Instruments
Ltd., Cambridge, Great Britain) was used in conjunction with the
electrochemical detector in some of the investigations.

    Sample introduction was by means of a Hamilton 701N syringe in
combination with a septum injector (HETP Components, Macclesfield,
Great Britain) or alternatively a Rheodyne valve injector with 20 µl
sample loop was used (HPLC Technology Ltd., Macclesfield, Great
Britain).

RESULTS AND DISCUSSION

Detection Systems

    (a) Electrode materials. The Bioanalytical Systems Inc. thin-
layer flow-cell (TL-4) was evaluated for the detection of 3,4-chloro-
fluoroaniline following chromatography of the compound on a
Lichrosorb RP-8 column with water:methanol (50 + 50) mobile phase
containing 0.1M potassium nitrate as electrolyte. Under these conditions a detection limit of 300 pg was obtained with longer term
drift being the major contribution to the noise of the system. The
carbon-paste working electrode tended to disintegrate when in contact
with mobile phases containing a significant proportion of methanol
(or less polar solvent), resulting in a continuously varying background current.

    The EDT Research wall-jet flow-cell LCO3 (see Figure 1) gave a
detection limit of 100 pg of 3,4-chlorofluoroaniline under similar
conditions. In operation the glassy carbon electrode showed much
better mechanical and chemical stability towards a wide range of
solvents of interest in HPLC than the carbon-paste electrode.

    Although these results indicate glassy carbon to be a better
electrode material when mixed solvent mobile phases are used, carbon
paste is clearly a useful electrode material in totally aqueous
mobile phases, as is evident from the work of Kissinger [5] and
others.

    (b) Modification of the LCO3 reference electrode. Initially
the LCO3 flow-cell gave a very noisy current signal when operated

with non-aqueous solvent systems. This was overcome by modification
of the reference electrode used in the flow-cell. The ceramic frit
was removed and a more porous glass frit was fitted. The filling
solution was then replaced with an agar gel saturated with potassium
chloride, which was squeezed through the glass frit to provide contact with the solution. This modified reference electrode provided a
lower resistance, less "noisy" contact with non-aqueous solvents and
was equally suitable for aqueous work.

The LCO3 flow-cell together with modified reference electrode
and glassy carbon working electrode was used in all the work reported
in the following sections of this paper. The more recent EDT
Research flow-cell LCA-13 has also been used in subsequent work and
shows a similar performance to the LCO3. The LCA-13 reference electrode appears to work in semi-aqueous (e.g., water:methanol, 50 + 50)
solvent systems with no increase in noise over that obtained with
aqueous solvent systems. It has not, however, been examined with
totally non-aqueous solvent systems which proved particularly "noisy"
with the original unmodified LCO3 reference electrode.

Determination of Electroactivity

(a) Standard technique. The determination of electroactivity
(if any) of a compound and the necessary conditions for that activity, viz. potential, mobile phase, electrolyte and electrode
material, is a necessary pre-requisite for setting up an HPLC system
with electrochemical detection. Cyclic voltammetry is a suitable
technique for such studies and yields information on the electroactivity of compounds more quickly than the stepwise approach adopted
by some workers [6]. A number of working electrodes can be employed
although carbon paste is generally the most useful for these studies
of organic compounds since the current peaks are normally well-
resolved. The electrode surface is replaced with every fresh solution to avoid problems of surface disintegration or interference
from adsorbed films. From the cyclic voltammetry data the peak
potential will indicate the potential to which the detector should be
set. Some final stepwise adjustment of the potential on successive
injections is usually necessary to obtain the optimum sensitivity.
The differences between the diffusion-controlled conditions in cyclic
voltammetry and the hydrodynamic conditions in HPLC detection frequently result, in the case of oxidations, in the optimum potential
being shifted slightly to more positive values. Anderson et al. [7],
have attempted to quantify the difference between the optimum potential for operation under hydrodynamic conditions and the peak and
half-peak potentials obtained from cyclic voltammetry. Although this
may have application where well-resolved cyclic voltammograms are
obtained, in the authors' experience many compounds that are less
easily oxidized do not show well-resolved cyclic voltammograms at
glassy carbon. (The method requires that the same electrode material

be used for both the cyclic voltammetry studies and the HPLC detection). In any case the detector potential should be set at the
minimum value necessary to yield the maximum peak height for the
chromatographed component of interest. This minimizes the background
current and its associated noise.

   (b) Microscale technique. A novel microscale voltammetric
technique has been developed.[8] The technique involves the deposition of analyte solution (100-500 nl) on the surface of the electrode followed by evaporation of the sample solvent. The electrode
is then placed in a blank electrolyte solution together with reference and counter electrodes, and the potential is scanned and the
current monitored using conventional instrumentation. In this way it
is possible to measure the electrochemical behavior of compounds
confined to the surface. It is also possible to use a second working
electrode with no deposited sample to monitor only the background
current, and addition of a simple subtracting circuit then allows
simultaneous background current correction to be implemented. Using
this latter approach well-resolved cyclic voltammograms can be obtained from less than 10 nanograms of material for a number of electroactive organic compounds. This technique enables scouting for
electroactivity prior to setting up an HPLC electrochemical detection
system to be implemented even when only small quantities of standards
are available.

Choice of Chromatographic Conditions

   The conditions required for electroactivity of a compound can
place constraints on the possible chromatographic methods that can be
used. In general solvents or mixtures of solvents polar enough to
dissolve suitable electrolytes must be used. This places restric-
tions on the use of solvents with polarities less than that of
methylene chloride. Thus if hexane is used as the solvent an excess
of a more polar solvent is required to provide sufficient polarity
for the dissolution of a suitable electrolyte. In practice this
limits the use of adsorption or normal phase partition chromatography
to those cases where more polar solvents or mixtures can be used.
Reverse-phase partition, ion-pair, ion-exchange and size exclusion
are modes of separation which are more compatible with electrochemical detection. An electrolyte can be dissolved in the mobile
phase normally without causing any deleterious effects on the
chromatographic separation. The use of non-aqueous reverse-phase
chromatography [9] is of particular interest for the exploitation of
electrochemical detection since the solvents used in such separations
often allow greater potentials to be employed, thereby extending the
range of application of the technique. This approach has been
exploited in some of the separations outlined in this paper.

   Table 1 lists the accessible potentials for some of the solvents
commonly used in reversed-phase chromatography. The values obtained

in practice for a given solvent are critically dependent on the solvent purity, electrode material and electrolyte. Table 2 lists some of the more common electrolytes together with their decomposition potentials.

Using cyclic voltammetric data, existing HPLC information and the tables below, a chromatographic system can often be developed which meets the requirements of both separation and detection.

Table 1.   Accessible Potential Range for Some Solvents

Solvent	Cathodic limit V vs SCE	Anodic limit V vs SCE	Reference
Water	−2.7	1.5	10
Methanol	−2.2	1.8	11
Acetonitrile	−3.5	2.4	12
Dimethylformamide	−3.5	1.5	13
Tetrahydrofuran	−3.6	1.8	14
Methylene dichloride	−1.7	1.8	15
Acetic acid	−1.7	2.0	16, 17

SCE = saturated calomel electrode
The above figures can only be taken as a guide since the actual potential limit is dependent on solvent purity, electrode material and electrolyte.

Table 2.   Accessible Potential Ranges in Dimethyl-sulfoxide Containing Different Supporting Electrolytes (0.1M)[18]

Electrolyte	Cathodic limit V vs SCE	Anodic limit V vs SCE
$LiClO_4$	−2.68	2.10
$KClO_4$	−2.33	2.10
$NaClO_4$	−2.08	2.10
KNO_3	−2.33	2.10
KBF_4	−2.33	2.10
$K_2S_2O_8$	−2.33	2.10
$LiCl$	−2.68	1.52
Me_4NCl	−2.40	1.52
Et_4NClO_4	−2.30	2.10
Bu_4NBr	−2.40	1.45

SCE = saturated calomel electrode.

Analytical Parameters

(a) Linear range. A linear dynamic range for anthracene from
<2.5 ng to 1 µg was obtained. Some curvature was noted for amounts
greater than 1 µg, possibly resulting from non-linearity of the LC-2A
electronics at high oxidation currents. Other compounds examined
such as aromatic amines and amino acids showed a similar or greater
linear dynamic range. A linear range of 10^3 is useful for most
analytical applications and is similar to that reported by other
workers.[1]

(b) Effect of electrolyte concentration. Electrolyte is added
to the mobile phase to ensure the conductivity necessary for electrochemical reaction. In all experiments the electrolyte was added to
the mobile phase prior to chromatographic separation. Comparison of
chromatograms obtained with and without electrolyte dissolved in the
mobile phase, using UV detection to monitor the separation, indicated
that addition of electrolyte generally had no adverse effects on the
chromatography.

The effect of the variation of electrolyte concentration on
sensitivity was investigated for cholesterol under conditions given
in Table 4. For the range 20-100 mM of sodium perchlorate in
acetonitrile there was little variation in the sensitivity or noise.
Concentrations of electrolyte in the middle of this range were used
for most of this work.

(c) Precision. The precision was measured as the relative
standard deviation (rsd) of a repeated injection of the same quantity
of sample. Using a second detector (UV), whose precision was more
accurately known, in series with the electrochemical detector, it was
found necessary to thermostat the column and electrochemical detector
in order to obtain the best precision. The precision of the electrochemical detector using 250 ng injections of anthracene was found to
be 1.5% rsd compared with 1.1% rsd for the UV detector for the same
sequence of injections. Although the figure for the electrochemical
detector is slightly worse than for the UV detector for the majority
of trace level determinations this precision will be perfectly
adequate. Using an automated sample injector it should be possible
to improve the precision figures for both detection systems.

(d) Effect of temperature stabilization. When the electrochemical detector is operated at high sensitivity fluctuations in the
detector/mobile phase temperature cause fluctuations in the standing
current. This feature is usually manifest as a sloping or drifting
baseline together with shorter-term noise. Some of the shorter-term
or high frequency noise can be removed by low pass filtering,
although with most systems this is restricted by the concomitant rise
in time constant. Time constant values greater than one second
cannot normally be used on most HPLC separations unless the retention
time is greater than 10 minutes.

Table 3. The Effect of Temperature Stabilisation on
Detector Drift and Noise

	No stabilisation	With stabilisation
Drift	1 nA/h	250 pA/h
Noise	40 pA (peak to peak)	20 pA (peak to peak)

Placing the electrochemical detector in a temperature-controlled environment, preferably together with sample injector and column, can improve the noise and drift of the system as shown in Table 3.

A fourfold reduction in baseline drift and twofold reduction in noise was found when temperature stabilization was implemented. This is a much smaller improvement than that reported by Purnell and Warwick[19], although their noise and drift before stabilization were at least a factor of ten worse than the comparable figures obtained in this study.

The actual values of noise and drift will depend on the particular system and applied potential, but the improvement with temperature stabilization is usually similar, ie. a fourfold improvement in drift, and is worth having if trace level work is to be attempted.

Applications

(a) Aromatic amines. The electrochemical oxidation of aromatic amines is well established [20,21] and this can be exploited in the detection of these compounds following separation by HPLC. Figure 2 illustrates the detection of 3,4-chlorofluoroaniline. Using the conditions given in Figure 1 the detection limit for this compound is 100 pg at a signal:noise of 4:1.

(b) Organic acids. Carboxylic acids can be oxidized electrochemically via the Kolbe reaction mechanism.[20] It was found possible to detect several organic acids, although no study was undertaken to confirm that the acids investigated oxidized via the Kolbe reaction. Tetrabutylammonium fluoroborate was added to the mobile phase at a concentration of 20 mM. This served both as a supporting electrolyte and as an ion-pairing reagent and reasonable chromatographic efficiency was achieved as shown in Figures 3 and 4. Detection limits for both compounds (∿10 ng at a signal:noise of 4:1) are appreciably better than those obtainable with UV detection for these acids.

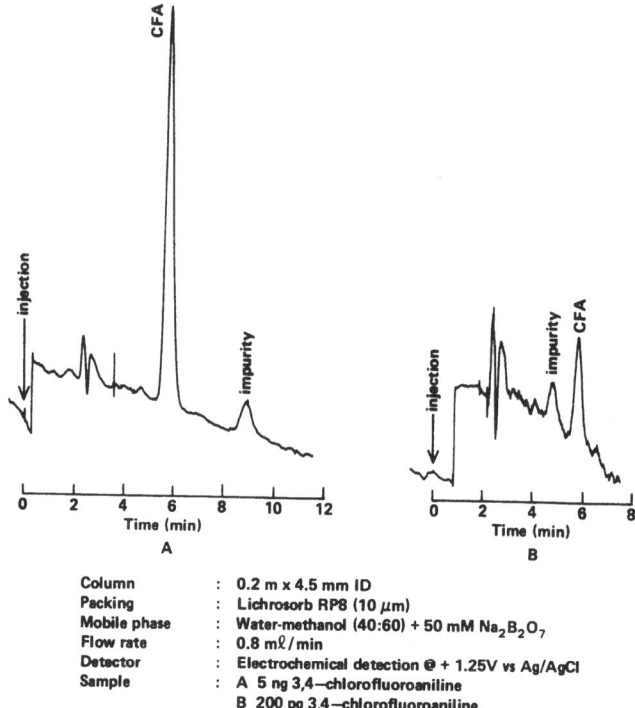

Column	:	0.2 m x 4.5 mm ID
Packing	:	Lichrosorb RP8 (10 μm)
Mobile phase	:	Water-methanol (40:60) + 50 mM Na$_2$B$_2$O$_7$
Flow rate	:	0.8 mℓ/min
Detector	:	Electrochemical detection @ + 1.25V vs Ag/AgCl
Sample	:	A 5 ng 3,4–chlorofluoroaniline
		B 200 pg 3,4–chlorofluoroaniline

Fig. 2.   Chromatograms of 3,4-chlorofluoroaniline (CFA).

(c) <u>Sterols</u>. Steroids containing phenolic moieties can be
oxidized electrochemically relatively easily but the electrochemical
detection of saturated sterols following separation by HPLC has not
been previously reported. A method was developed for the separation
of sterols by HPLC using a column packed with Hypersil ODS and
acetonitrile as the mobile phase. It was found that electrochemical
detection could be employed for these compounds if sodium perchlorate
was added to the mobile phase. Table 4 outlines the electrochemical
and chromatographic conditions and gives the detection limits obtained for some of these compounds. The use of a non-aqueous mobile
phase with a reverse-phase column to separate these rather non-polar
molecules permits the high applied potentials necessary for the
detection of these compounds. Figure 5 shows a separation of ergos-
terol, cholesterol, campesterol and β–sitosterol. Detection by UV
absorption yields detection limits of ca. 100 ng for some of these
sterols. Electrochemical detection offers an improvement in sensitivity by a factor of about ten in most cases.

(d) <u>Non-ionic surfactants</u>. Non-ionic surfactants are a group
of compounds that are often difficult both to chromatograph and to
detect (if they contain no UV chromophore) using HPLC techniques.

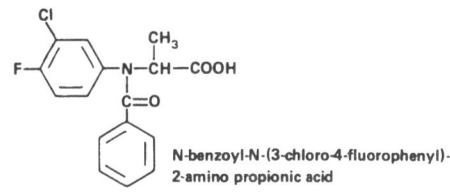

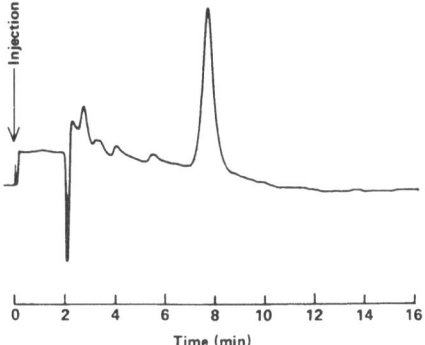

Column	:	0.2 m x 4.5 mm ID
Packing	:	Hypersil ODS (5.5 μm)
Mobile phase	:	Water-methanol (65:35) + 20 mM $(C_4H_9)_4NBF_4$
Flow rate	:	1.0 mℓ/min
Detector	:	Electrochemical detection @ + 1.94V vs Ag/Ag Cl
Sample	:	480 ng N-benzoyl-N-(3-chloro-4-fluorophenyl)-
		2-amino propionic acid

Fig. 3. Chromatogram of N-benzoyl-N-(3-chloro-4-fluorophenyl)
2-amino propionic acid.

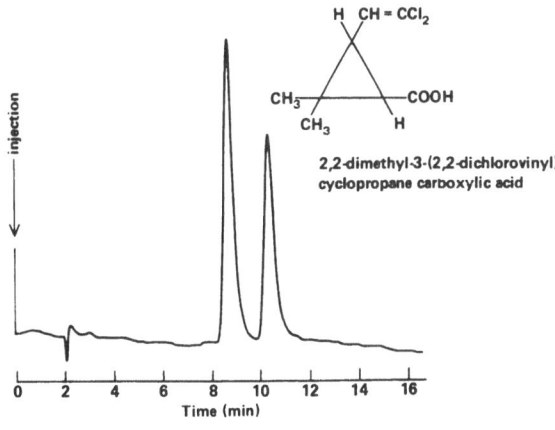

Column	:	0.2 m x 4.5 mm ID
Packing	:	Hypersil ODS (5.5 μm)
Mobile phase	:	Water-methanol (65:35) + 20 mM $(C_4H_9)_4NBF_4$
Flow rate	:	1.0 mℓ/min
Detector	:	Electrochemical detection @ + 1.94V vs Ag/Ag Cl
Sample	:	1 μg 2,2-dimethyl-3-(2,2-dichlorovinyl)
		cyclopropane carboxylic acid

Fig. 4. Chromatogram of 2,2-dimethyl-3(2,2-dichlorovinyl)
cyclopropane carboxylic acid showing isomer separation.

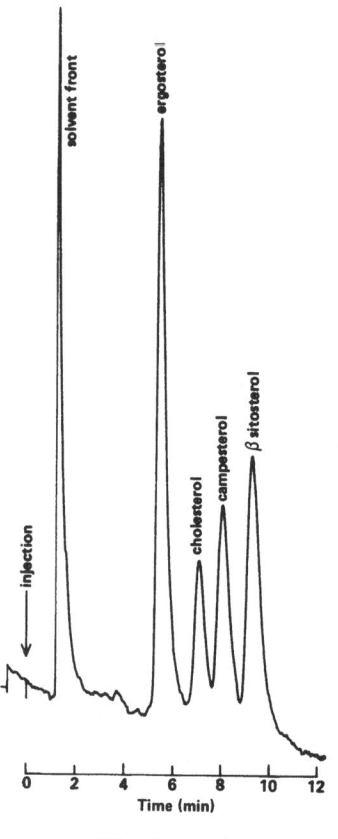

Column	:	0.2 m x 4.5 mm ID
Packing	:	Hypersil ODS (5.5 μm)
Mobile phase	:	Acetonitrile + 50 mM NaClO$_4$
Flow rate	:	2.1 mℓ/min
Detector	:	Electrochemical detection @ + 1.90V vs Ag/Ag Cl
Sample	:	1 μg ergosterol, 1 μg cholesterol
		2.5 μg campesterol and 2.5 μg β—sitosterol

Fig. 5.   Chromatogram of a mixture of sterols.

Electrochemical detection has been shown to be useful for several non-ionic surfactants and a summary of the results is given in Table 5.

The nonyl- and octyl-phenolethoxylate surfactants, including Triton-X 100, can be detected electrochemically. The exact mechanisms of the electrochemical reaction has not been studied, but is believed to involve the oxidation of the terminal hydroxyl group. Figure 6 illustrates the separation (into its oligomers) of a typical nonylphenolethoxylate, Lutensol AP-10, using electrochemical detection.

Table 4. Chromatographic and Electrochemical Data for
          Some Sterols

Compound	Retention time (min)	Oxidation potential V vs SCE	Detection limit (ng)
Cholesterol	6.1	1.70	40
Ergosterol	5.5	1.08	20
Lanosterol	6.0	1.38	10
Dihydrolanosterol	8.4	1.38	10
β-Sitosterol	8.8	1.78	100
Campesterol	7.6	1.78	100
Cholestanol	6.3	ND	500

ND = Not determined
Column - 0.20 m x 4.5 mm ID
Packing - Hypersil ODS (5.5 μm)
Mobile phase - Acetonitrile + 40 μM NaClO$_4$
Flow rate - 2.1 ml/min
Detector - Electrochemical detection at + 1.90 V vs Ag/AgCl

The other group of non-ionic surfactants studied were the
Pluronics. These are ethylene oxide/propylene oxide block co-
polymers and some of these have been chromatographed as a single peak
(no resolution of any oligomers) using size exclusion chromatography.
These compounds show weak UV absorption but can be detected electrochemically.

(e) Other compounds. Several other classes of compounds have
been briefly examined with some success. Anthracene can be detected
with good sensitivity yielding a detection limit of ca. 400 pg
(signal:noise of 4:1). The conditions used to chromatograph and
detect anthracene were: a Lichrosorb RP-8 column eluted with a
mixture of water:methanol (10 + 90) containing 50 mM potassium
nitrate with an applied potential of + 1.7 V vs Ag/AgCl. Other
polynuclear aromatic hydrocarbons can be oxidized electrochemically,
although for the most part at higher potentials, and would probably
be detected with comparable sensitivity.

Argine and valine can be detected electrochemically with a
detection limit for argine of ca. 100 ng (signal:noise of 4:1). The
detector is rather less sensitive towards valine. The conditions
used to chromatograph and detect these amino acids were: a Lichrosorb
RP-8 column eluted with 0.1M aqueous phosphate buffer (pH 6.7) with
an applied potential of +1.62 V vs Ag/AgCl. The literature contains
reports [22,23] of other amino acids being detected electrochemically
and this method may be generally applicable for amino acids.

Table 5. Summary of HPLC Conditions for the Determination of Some Surfactants Using Electrochemical Detection

Compound	Structure	Column packing	Mobile phase	Flow rate ml/min	Comments	Detection Limit µg
Cemulsols Lutensols	Nonylphenolethoxylates typically 1–30 ethoxy-groups	Partisil-10 PAC (10 µm)*	CH_3CN-CH_3OH (95:5) + 40 mM $NaClO_4$	1.6	Oligomer separation	ca. 1
Triton X series	Octylphenolethoxylates typically 1–30 ethoxy-groups	Partisil-10 PAC (10 µm)*	CH_3CN-CH_3OH (95:5) + 40 mM $NaClO_4$	1.6	Oligomer separation	ND
Pluronics	$HO(C_2H_4O)_a(C_3H_6O)_b(C_2H_4O)_aH$ MW 950–f4,000	Lichrosorb SI-60 (5 µm)*	CH_3OH-H_2O (65:35) + 50 mM $Na_2B_2O_7$	0.65	No oligomer separation	ca. 2
		Partisil-5 (5 µm)*	CH_3OH-H_2O (60:40) + 50 mM KNO_3 + 5 m$M_2Na_2B_7O$	0.35	No oligomer separation	ca. 2

*Column: 0.20 m x 3.5 mm ID
ND = Not determined

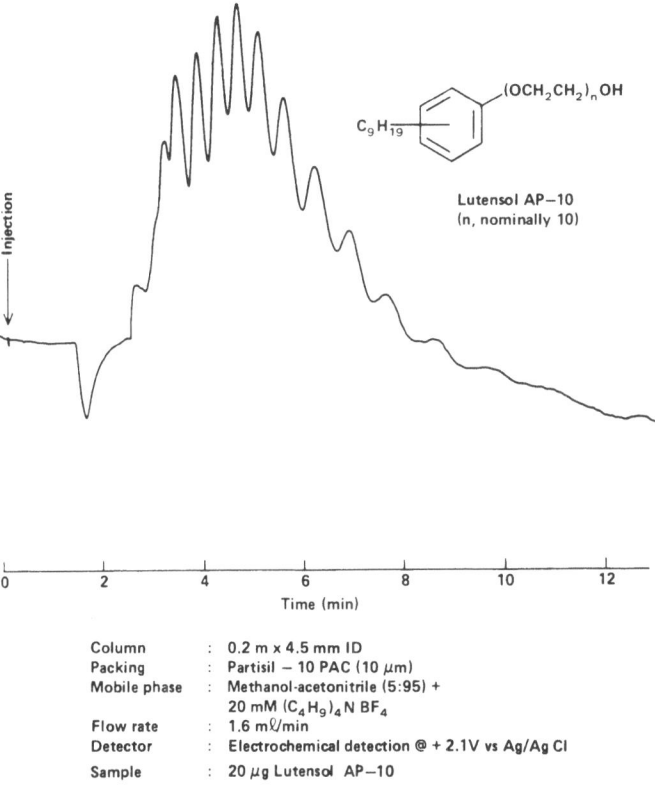

Fig. 6. Chromatogram of Lutensol AP-10 showing oligomer separation.

CONCLUSIONS

Electrochemical detection in HPLC has become established for some, more specialized, applications such as catecholamine analysis though it can be exploited for a far wider range of compounds. The work in this paper has attempted to investigate some of the basic properties of an electrochemical detection system and some more difficult applications. The detector has a linear dynamic range and precision that are comparable with those of other detectors for HPLC. It is, however, more dependent on temperature than, for example, the UV absorption detector and must be operated in a temperature controlled environment to obtain the lowest detection limits. For many electroactive compounds with moderate oxidation potentials, the electrochemical detector can yield sub-nanogram detection limits. Even for compounds with high oxidation potentials, in many cases the electrochemical detector has significantly greater sensitivity than the UV detector.

Electrochemical detection can be employed with most chromatographic modes in HPLC although polar mobile phases containing dissolved electrolytes must be used for electrochemical detection to operate. This work has demonstrated that it is possible to use electrochemical detection with non-aqueous solvents, such as acetonitrile, thereby expanding the areas of possible use of the technique. The application of the detector for monitoring sterols, organic acids and some non-ionic surfactants, in many cases using non-aqueous mobile phases, has been demonstrated. For many of these compounds electrochemical detection offers lower detection limits than any other direct detection system.

Future developments of electrochemical detector cells and applications to a wider range of compounds can be foreseen and could lead to greater exploitation of these detectors. /There is still scope for improvement in cell design, particularly in the improvement in flow patterns to reduce noise./ Novel electrode materials could be profitably exploited, for example, if electrocatalysis towards a given substrate could be demonstrated or low background currents achieved. Chemical derivatization has been frequently exploited in polarography to expand its range of applicability and similar techniques, particularly if implemented post-column and in-line, could likewise expand the areas of application of electrochemical detection in HPLC.

REFERENCES

1. K. Stulik and V. Pacakova, J.Electroanal.Chem., 129:1 (1981).
2. P. T. Kissinger, C. S. Bruntlett, and R. E. Shoup, Life Sciences, 28:455 (1981).
3. W. R. Smyth and C. G. Frischkorn, Fresenius Z.Anal.Chem., 301:220 (1980).
4. T. J. N. Webber and E. H. McKerrell, J.Chromatog., 122:243 (1976).
5. L. J. Felice, W. P. King, and P. T. Kissinger, J.Agric.Food Chem., 24:380 (1976).
6. J. W. Dieker, W. E. van der Linden, and H. Poppe, Talanta, 26:511 (1979).
7. J. L. Anderson, D. E. Weisshaar, and D. E. Tallman, Anal.Chem., 53:908 (1981).
8. A. J. Samuel and T. J. N. Webber, J.Electroanal.Chem., 131:391 (1982).
9. N. A. Parris, J.Chromatog., 157:161 (1978).
10. M. M. Baizer, ed., Organic Electrochemistry, Dekker 210 (1973).
11. W. Roger and S. M. Kipnes, Anal.Chem., 27:1916 (1955).
12. J. P. Billon, J.Electroanal.Chem., 1:486 (1960).
13. M. Breant, M. Bazoin, C. Buisson, M. Dupin, and J. M. Rebattu, Bull.Soc.Chim.Fr., 5065 (1968).
14. J. Perichon and R. Buvet, Electrochim.Acta, 9:567 (1964).
15. A. J. Bard, ed., Electroanalytical Chemistry, 123 (1969).

16. L. Eberson and K. Nyberg, J.Am.Chem.Soc., 88:1686 (1966).
17. G. B. Bachmann and M. J. Astle, J.Am.Chem.Soc., 64:1303 (1942).
18. J. Courtot-Coupez and M. Le Demezet, Bull.Soc.Chim.Fr., 4744 (1967).
19. C. J. Purnell and C. J. Warwick, Anal.Proc., 151 April (1981).
20. R. N. Adams, Electrochemistry at solid electrodes, Dekker 372 (1969).
21. J. P. Hart, M. R. Smyth, and W. F. Smyth, Analyst, 106:146 (1981).
22. Y. Takata and G. Muto, Anal.Chem., 48:1864 (1973).
23. B. Fleet and C. J. Little, J.Chromatog.Sci., 12:747 (1974).

APPLICATIONS OF HIGH PERFORMANCE LIQUID CHROMATOGRAPHY WITH

ELECTROCHEMICAL DETECTION IN CLINICAL CHEMISTRY

Douglas A. Richards

Department of Biochemistry
Royal Sussex County Hospital, Eastern Road
Brighton, Sussex (Gt.Britain)

SUMMARY

    The value of electrochemical detection following high perform-
ance liquid chromatography is compared with that of the more widely
used optical methods of detection. Its advantages are illustrated by
its application to three areas of clinical chemistry that had previously posed analytical problems. Its sensitivity allowed its use
for the measurement of plasma catecholamines. Its selectivity was
employed to produce rapid methods for the determination of urinary
levels of catecholamine and tryptophan metabolites. Finally, its
value for the estimation of urinary oxalic acid is shown. Future
developments such as increasing the range of detectable compounds by
derivatization are briefly discussed.

INTRODUCTION

    It is only during the last 5 years that high performance liquid
chromatography (HPLC) has been used to any extent in clinical chemistry and yet, in that short time, it has already been established as
a widely used analytical technique and is continuing to grow in
popularity. Early problems with the equipment, particularly high
pressure pumps, have largely been solved so that there is greater
reliability than in the earlier years. However, the major limitation
of analysis by HPLC lies in the detection systems. By far the most
commonly used detector is the UV spectrophotometer, preferably with a
variable-wavelength facility. However, this is not sensitive enough
for the more demanding clinical applications, and many compounds of
interest do not contain UV-absorbing chromophores. Increased sensi-

tivity can be obtained by using fluorimetric detectors, measuring the
fluorescence either of the compound or of a fluorescent derivative.

Whilst these two systems cover the majority of clinical applications of HPLC, there has always been room for other means of detection. The two main advantages of electrochemical detection (ECD) are
its sensitivity and its greater selectivity. For many compounds ECD
is more sensitive than UV or fluorimetric detection and this is
particularly valuable in clinical chemistry where the measurement of
picogram quantities is frequently required. Selectivity is achieved
by the choice of appropriate operating potentials which allow the
detection of compounds of interest whilst other electroactive species
requiring a higher potential are not "seen". Even at the upper limit
of the available potential range, the electroactive compounds in
biological material are far less in number than those which absorb UV
light. This means that samples such as blood, serum or urine require
less pre-treatment before injection into the chromatographic system
if ECD is used. However, many compounds of interest are not electroactive or else produce very small current signals. In addition there
are also substances with which the electrochemical reaction gives
insoluble products. These products build up on the working electrode
surface causing lowered response to subsequent samples. Finally,
electrochemical detectors are, as yet, not as reliable as optical
detection systems. They also require a certain amount of expertise
in their setting up in order to achieve stable performance over a
long period and even small amounts of gas in the mobile phase cause
instability. For these reasons ECD is not a superior technique to UV
or fluorimetric detection but rather a complementary system capable
of overcoming some of the inadequacies of the two major means of
detection.

SOME APPLICATIONS OF ELECTROCHEMICAL DETECTION IN CLINICAL CHEMISTRY

The study of catecholamines and their metabolites has grown
enormously in recent years since the advent of HPLC coupled to highly
sensitive detection systems. The catecholamines play a major role in
the function of the body's nervous system. Dopamine and noradrenaline exert a marked influence on the vascular system, whilst adrenaline, synthesized largely in the adrenal medulla, affects the rate of
many metabolic processes, particularly carbohydrate metabolism. The
major metabolic pathways of catecholamine inactivation are shown in
Figure 1. Abnormal levels of these amines and their metabolites have
been associated with a number of disease states, notably Parkinson's
disease [1], neural tumours such as phaeochromocytoma[2] and hyper-
tension [3].

In the past, urinary catecholamines have been determined fluori-
metrically after ion-exchange chromatography and conversion to the
highly fluorescent trihydroxy indole derivatives [4]. However, this

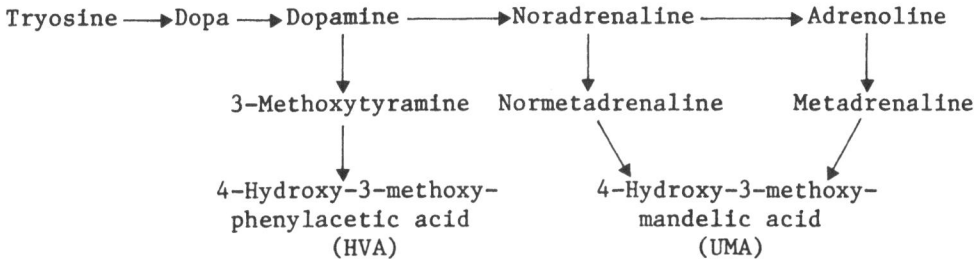

Fig. 1. The major pathways of catecholamine metabolism.

method was not sufficiently sensitive to determine the picogram
quantities present in plasma, an assay considered by many to be the
only reliable index of catecholamine metabolism. Until HPLC became
available, plasma catecholamines had usually been determined by
radio-enzymatic procedures which are costly, time consuming, unre-
liable and beyond the capacity of most clinical chemistry laboratories. However, catecholamines can readily be separated by HPLC
using ion-exchange or reversed-phase columns and they can be ex-
tracted from plasma with acceptable recoveries. The difficult part
of the assay is the detection. Although they absorb UV light, UV
detection is not sensitive enough for catecholamine levels. Similarly, their weak native fluorescence cannot be used. This leaves
two alternatives: fluorimetric detection following post-column
derivatization of the amines[5] or ECD [6,7].

ECD has comparable sensitivity with that of post-column derivatization using the trihydroxy indole reaction but has the advantage
of not requiring a post-column reaction system which introduces some
peak broadening. In addition, catecholamines are oxidized at a
relatively low potential, and so selectivity is good.

In our laboratory we have used both of these detection systems
for plasma catecholamines, and, in spite of these advantages of ECD,
its instability and unreliability when operated at its sensitivity
limits have led us to choose fluorimetric detection for routine
purposes at present.

The diagnosis of neural tumours such as phaeochromocytoma and
neuroblastoma is heavily dependent on the laboratory measurement of
urinary levels of methylated catecholamines and of their two acidic
catabolites, 4-hydroxy-3-methoxymandelic acid (VMA) and 4-hydroxy-3-
methoxyphenyl acetic acid (HVA).

The methylated catecholamines have most commonly been estimated
by spectrophotometric determination of the vanillin formed from their
oxidation by periodate [8], after extraction from the urine by ion-
exchange chromatography. A method using HPLC and ECD has now been

described for these substances [9]. In this instance ECD is used not because of its sensitivity, since the urinary levels of methylated catecholamines are sufficient to allow the use of UV detection, but because of its selectivity which permits sample preparation stages prior to injection to be kept to a minimum.

VMA is quantitatively the most important of the catecholamine metabolites. There have been two approaches to its measurement in urine. The method of Pisano et al.[10] employs oxidation to vanillin following a lengthy extraction procedure. Whilst this is fairly specific, it is too tedious for the large numbers of VMA analyses handled by a clinical laboratory. Colorimetric methods following reaction with p-nitroaniline[11,12] are much quicker but sacrifice a certain degree of specificity. In a recent paper, a method for VMA in urine employing HPLC and ECD[13] has been shown to overcome these problems. A rapid and simple extraction procedure is used to produce a buffered extract of organic acids. The selectivity of ECD is then utilized so that relatively few peaks, including VMA, appear on the final chromatogram.

The same workers have also described a method for HVA [14]. This substance is of particular value in the diagnosis and treatment of neuroblastomas in children. The extraction procedure is the same as for VMA, the only difference in the method being the composition of the mobile phase used for the chromatography in order to elute the HVA earlier. In our laboratory we have developed a method which allows the simultaneous determination of VMA and HVA in urine; a chromatogram of this is shown in Figure 2.

Another area of clinical interest which has benefited from the use of ECD is the study of tryptophan metabolism. Tryptophan is an essential amino acid which can be metabolized in a number of ways as illustrated in Figure 3. One of the minor pathways leads to the synthesis of the neurotransmitter 5-hydroxytryptamine (5HT), abnormal brain levels of which have been associated with a number of diseases including Down's syndrome[15] and depression [16]. As part of a study into the biochemistry of depressive illnesses, I investigated methods for the determination of tryptophan metabolites in blood and urine. HPLC proved to be a useful technique for separating these compounds, so the possibility of using ECD was examined [17]. Of 29 metabolites tested, 26 were electroactive. Each substance was injected repeatedly, the operating potential of the cell being increased in 100 mV increments and the height of the resultant peak being measured. A polarographic curve was plotted for each compound from which the half-wave potential $E_{\frac{1}{2}}$ could be calculated. Figure 4 shows the curve obtained for 5-hydroxyindole acetic acid (5HIAA). The $E_{\frac{1}{2}}$ values for the 26 electroactive compounds are shown in Table 1, and from this it can be seen that, by selection of the appropriate operating potential, only those compounds from the pathway leading to the formation and excretion of 5HT are detected. This is of particu-

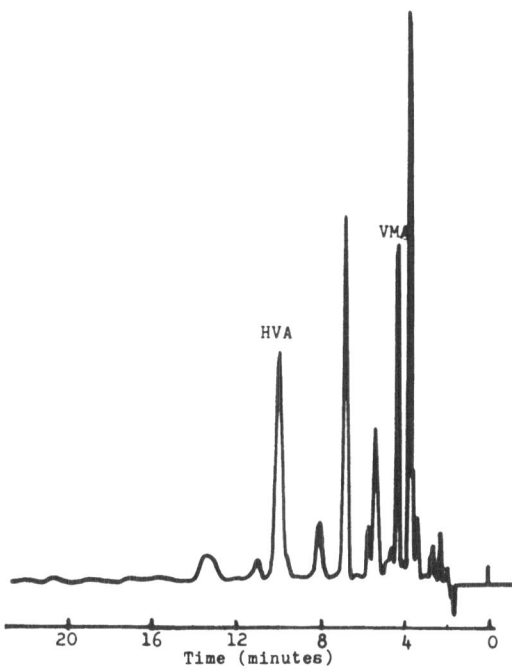

Fig. 2. Separation of urinary organic acids on a 120 mm x 5 mm
column of ODS-Hypersil (the mobile phase composition was 20%
methanol, 0.4% sodium nitrate, 0.04% sulphuric acid and
0.012% sodium lauryl sulphate; detector operating potential,
+1.0 V; sensitivity, 300 nA).

lar value in the study of disturbances of 5HT metabolism and is a
unique property of this type of detector. It was also shown that the
detector response was linear over a wide range of concentration and
that, under the conditions used, the limit of detection for each
metabolite was 2 pmol.

Another, quite different, example of the value of ECD in helping
to solve a difficult methodological problem is the measurement of
urinary oxalate. Excessive excretion of oxalate in the urine is
likely to lead to the formation of calculi in the kidney or the
urinary tract. By maintaining urinary oxalate excretion within
normal limits, the likelihood of oxalate stone formation is considerably reduced and thus this test is frequently requested by urol-
ogists.

There have been two main approaches to this assay. The first is
to precipitate the oxalic acid as its insoluble salt, calcium oxalate, over a period of 48 h and then to determine the concentration
of this colorimetrically after reduction to glycollic acid, conver-

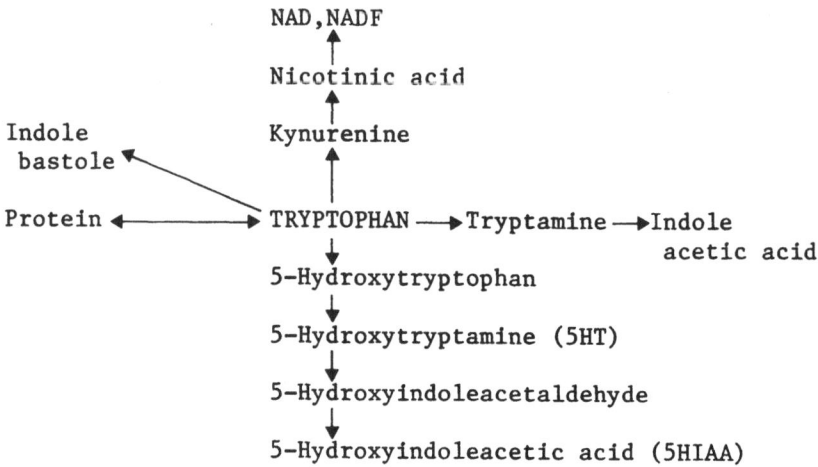

Fig. 3.   The main metabolic pathways of tryptophan.

sion to formaldehyde and formation of a colored complex with chromotropic acid [18]. Apart from being so time consuming, this method also requires the handling of boiling concentrated acids.   An alternative is the use of enzymes[19] such as oxalate decarboxylase. Enzymic methods have the advantage that the precipitation stage can be eliminated but they are far too expensive for routine purposes.

It was noted in our laboratory that oxalic acid underwent electrochemical oxidation and this was used as the basis of a new method for its determination.   The oxalic acid is precipitated as calcium oxalate, which is dissolved in acid and detected electrochemically after HPLC.   Oxalic acid requires a relatively high operating potential (+1.6 V) for oxidation in our system, but this presents no selectivity problems since, by careful washing of the precipitate, only a single peak is seen on the chromatogram at the sensitivity required.   A similar method has been described by Mayer et al. [20,21]; this also requires a precipitation step although the step has been shortened.   If the precipitation stage, with its inherent loss of recovery, can be eliminated and replaced with solvent extraction, the method can be further simplified and we are currently investigating this.

HPLC and ECD have been used for other purposes in clinical laboratories, e.g. the measurement of uric acid[22] and thyroid hormones [23].   However, as these assays are commonly requested in large numbers and are handled with well-established batch techniques which are frequently automated, it is most unlikely that HPLC and ECD will ever be used for routine measurement of these substances.

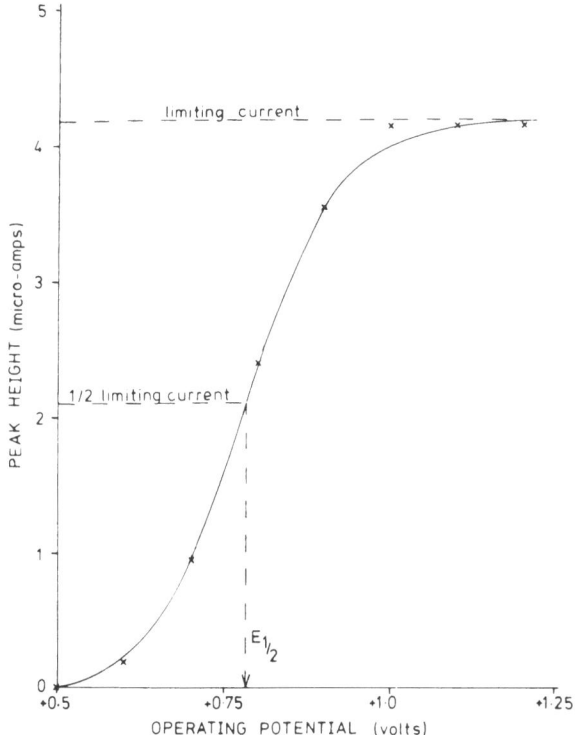

Fig. 4.   Polarographic curves for 5HIAA. The half-wave potential
          $E_{\frac{1}{2}}$ is determined as that voltage at which the current
          produced by the oxidation is half of the limiting current.

     Looking ahead to what lies in the future of ECD in clinical
chemistry, it is probable that the next major advance will be in the
formation of electroactive derivatives of non-electroactive compounds
to enlarge the range of substances that can be detected. A recent
example of this is the determination of ketosteroids following derivatization[24]. An even more novel approach to expanding the range
of detectable compounds is the use of electrogenerated reagents [25].
Using electrogenerated bromine, introduced into the chromatography
column effluent, to react with eluted compounds, nanogram quantities
of un-derivatized fatty acids, prostaglandins and phenols have been
detected.

CONCLUSIONS

     The value of ECD as a detection technique has been illustrated
by the preceding examples. For plasma catecholamines it has been

Table 1. Half-wave Potentials $E_{\frac{1}{2}}$ for Tryptophan

Compound	$E_{\frac{1}{2}}$ (V)
5-Hydroxyindole	+0.80
5-Hydroxyindole-3-acetic acid	+0.80
5-Hydroxytryptophol	+0.80
N-Acetyl-5-hydroxytryptamine	+0.80
5-Hydroxytryptamine	+0.85
5-Hydroxytryptophan	+0.85
Indoxyl sulfate	+1.10
Tryptophol	+1.15
5-Methoxytryptophol	+1.15
3-Methylindole	+1.15
N-Methyltryptamine	+1.15
Indole-3-acrylic acid	+1.15
Indole-3-acetic acid	+1.15
Indole-3-pyruvic acid	+1.15
Indole-3-propionic acid	+1.15
5-Methoxyindole-3-acetic acid	+1.15
Indole-3-lactic acid	+1.20
N-Acetyltryptophan	+1.20
5-Methoxytryptamine	+1.20
Tryptophan	+1.25
Tryptamine	+1.25
Indole-3-acetamide	+1.25
Indole-3-acetonitrile	+1.25
Indole-3-acetaldehyde	+1.25
Indoxyl acetate	+1.25
Indole	+1.30

used for its sensitivity. For the determination of the urinary methylated amines MA and HVA, ECD is the detection system of choice because its greater selectivity allows the use of less rigorous extraction procedures prior to injection. In the study of tryptophan metabolites, compounds of the important tryptophan hydroxylase pathway can be selectively detected. Finally, oxalic acid cannot be detected by UV or fluorimetric methods but it can be oxidized electrochemically.

REFERENCES

1. A. Barbeau, Can.Med.Assoc.J., 87:802 (1962).
2. E. L. Bravo, R. C. Tarazi, R. W. Gifford, and B. H. Stewart, N.Engl.J.Med., 301:682 (1979).
3. P. Weidmann, D. Hirsch, C. Beretta-Piccoli, F. C. Reubi, and W. H. Zeigler, Am.J.Med., 62:209 (1977).

4. H. Weil-Malherbe and A. D. Bone, J.Clin.Pathol., 10:138 (1957).
5. L. M. Nelson and M. Carruthers, J.Chromatogr., 183:295 (1980).
6. H. Hallman, L. Farnebo, B. Hamberger, and G. Jonsson, Life.Sci., 23:1049 (1978).
7. G. C. Davis, P. T. Kissinger, and R. E. Shoup, Anal.Chem., 53:156 (1981).
8. J. J. Pisano, Clin.Chim.Acta, 5:406 (1960).
9. R. E. Shoup and P. T. Kissinger, Clin.Chem., (Winston-Salem,NC), 23:1268 (1977).
10. J. J. Pisano, J. R. Crout, and D. Abraham, Clin.Chim.Acta, 7:285 (1962).
11. D. A. Richards, Med.Lab.Sci., 33:155 (1976).
12. D. A. Richards, Med.Lab.Sci., 38:111 (1981).
13. J. L. Morrisey and Z. K. Shihabi, Clin.Chem., 25:2043 (1979).
14. J. L. Morrisey and Z. K. Shihabi, Clin.Chem., 25:2045 (1979).
15. D. O'Brien and A. Groshek, Arch.Dis.Child., 37:17 (1962).
16. A. J. Coppen, J.Psychiatr.Res., 9:163 (1972).
17. D. A. Richards, J.Chromatogr., 175:293 (1979).
18. A. Hodgkinson and A. Williams, Clin.Chim.Acta, 36:127 (1972).
19. G. Kohlbecker, L Richter, and M. Butz, Clin.Biochem., 17:309 (1979).
20. W. J. Mayer and M. S. Greenberg, J.Chromatogr.Sci., 17:614 (1979).
21. W. J. Mayer, J. P. McCarthy, and M. S. Greenberg, J.Chromatogr. Sci., 16:656.
22. L. A. Pachla and P. T. Kissinger, Clin.Chim.Acta, 59:309 (1975).
23. B. R. Hepler, S. G. Weber, and W. C. Purdy, Anal.Chim.Acta, 113:269 (1980).
24. K. Shimada, M. Tanaka, and T. Nambara, Anal.Lett., 25:567 (1980).
25. W. P. King and P. T. Kissinger, Clin.Chem., 26:1484 (1980).

OPTIMIZATION OF AN ELECTROCHEMICAL DETECTOR USING A STATIC MERCURY
DROP ELECTRODE IN HIGH-PERFORMANCE LIQUID CHROMATOGRAPHY. ANALYSIS
OF THE ANTICANCER AGENT MITOMYCIN C IN PLASMA

W. J. van Oort, J. den Hartigh and R. J. Driebergen

Pharmaceutical Laboratory
Department of Analytical Pharmacy
Catharijnesingel 60, 3511 GH Utrecht
The Netherlands

ABSTRACT

    An electrochemical detector using a static mercury drop elec-
trode has been optimized for combination with high-performance liquid
chromatography. Parameters like pump noise, oxygen in mobile phase
and sample solution, nozzle, flow rate, working potential in the d.c.
- and the d.p.p.- mode, have been examined. The application of the
method in routine analysis has been illustrated by the analysis of
the anticancer agent mitomycin C in plasma.

INTRODUCTION

    High-performance liquid chromatography especially the reversed-
phase mode is widely used for separation of lipophilic, non-volatile
compounds in complex matrices. However, the sensitivity and the
selectivity of the detection methods may be improved, when compared
with some detection methods in gas chromatographic procedures. It is
obviously easier to detect compounds in a nearly empty stream such as
a gas stream, than to determine compounds in a bulky stream of a
solution. Electrochemical processes seem to be very promising for
the analysis of compounds in a stream of solute, as these methods
need an excess of ions in the solution to facilitate the electrochemical reactions. With respect to other detection methods, such as
ultra-violet and fluorimetric spectrometry, a completely different
type of selectivity can be achieved by electrochemical reactions.
The sensitivity depends on the specific characteristics of compounds
and electrodes, but has proved to be much higher in flow-cells than
in batch-electrochemical processes. The higher molar sensitivity can

71

be achieved by the very thin diffusion-layer, in thin-layer cells, as
well as in wall-jet cells. The apparent small volume of the measuring cells contributes to the small, absolute amount of compound to be
measured. Electrochemical detectors operated in the oxidative mode
using either (glassy) carbon, or gold electrodes can be used on a
routine base for detection of compounds containing, e.g., phenolic,
aromatic amine, methoxy and sulfur groups[1-3]. Reduction processes
at carbon electrodes can only be performed with compounds with low
reduction potential, like quinones, nitro- and nitroso-compounds
[4,5]. The high overvoltage of hydrogen and the continuously renewable fresh surface of a dropping mercury electrode (DME) have led to
several attempts to develop a detector based on the reduction of
compounds at a DME. Several types of cells have been developed.
Hanekamp et al. compared different designs of DME-based detector
cells[6]. Horizontal and vertical capillaries, normal and short drop
times, flow-streams reaching the DME vertically or horizontally and
the apparent cell-volume are parameters which are investigated extensively[7-13]. Most detectors are laboratory-made by electroanalytical specialists. Application in routine and control laboratories, mostly without the presence of an electrochemist, often leads
to discouraging results. The recent introduction of a static mercury
drop electrode assembly (SMDE), with a module for detection in flowing streams, promised a solution to many practical problems. However, even with the commercially available apparatus, routine analysis at trace level proved to be troublesome. Bond and Jones[14]
investigated the analytical performance of the static mercury drop
electrode in batch polarography and established remarkable improvements in the sensitivity of current-sampled d.c. - polarography at
the SMDE, compared with the expanding dropping mercury electrode
(DME). They postulated a converging tendency of the analytical
advantages of d.c., normal pulse and differential pulse polarography
at the SMDE. Schieffer[15] applied the SMDE-detector in a stability-
indicating HPLC-assay of diacetolol, and could determine 400 ng
diacetolol and degradation products with d.c. - polarography and 10
ng with d.p.p. No special precautions had been taken to eliminate
the interference of oxygen and pump noise. Samuelson et al. [16,17]
determined N-nitrosamines by SMDE-detection after HPLC-separation
with direct-current, normal pulse, differential pulse and square-wave
voltammetry. A detection limit of 0.8 ng was calculated by d.p.p.,
only three times lower than d.c., in agreement with the findings of
Bond and Jones[14]. They used a home-made pulse dampener, but pump
pulsations could not be fully suppressed. Argon gas was used for
removal of oxygen from sample and mobile phase. Vohra and Harrington
[18] found for n-nitrosamines the same detection limit with d.p.p.,
but a 30 times lower sensitivity with d.c.-polarography. Helium gas
was used for degassing. For the analysis of isosorbidedinitrate
Persson and Rosén[19] didn't remove oxygen by degassing but supplied
sulphite ions to the mobile phase. The relatively high detection
limit of 100 ng, obtained by d.c.- polarography, is due to diffusion
of traces of oxygen through plastic tubes. Bond et al. [20] used a

sulphite bed in the flow stream to remove oxygen. Hackman and Brooks [21] implanted several devices to remove oxygen and to decrease pump noise for the determination of chlordiazepoxide and metabolites, like a partly filled column, connected by a T-piece with the flow stream, degassing with helium, replacement of teflon tubes by special tubes, impervious to oxygen and insertion of a noise filter with a time constant of 3 s to diminish pump noise. The most important improvement was obtained by stabilising the potential of the reference electrode.

In this paper further optimisation of the HPLC-system with the SMDE-detector will be reported. As a test compound the anticancer agent mitomycin C has been used, containing an easily reducible quinone-structure. For this compound, an HPLC-assay in plasma with U.V-detection has already been worked out[22], while the electrochemistry has been studied extensively[23].

EXPERIMENTAL

Instrumentation

The HPLC-system consisted of a Model 6000A pumping system, a Model U6K loop injector (all from Waters Assoc., Milford, MA, USA). A Model 310 polarographic detector controlled by a Model 174 polarographic analyzer (EG & G, Princeton Applied Research Corp., Princeton, NJ, USA) was used. The electrochemical detector was operated either in the sampled direct-current mode or in the differential pulse mode. The reference electrode was a Ag/AgCl electrode; filling solution consisted of saturated AgCl in 3 M KCl. The drop size was 'large'. The column used was a 30 cm x 3.9 I.D. stainless-steel prepacked reversed-phase column containing 10 μm μBondapak C_{18} particles (Waters Assoc., Milford, MA, USA). The mobile phase consisted of mixtures of phosphate buffer pH 8.0 and methanol, 0.01 M Na_2SO_3. The stock solution of the mobile phase was deaerated continuously by nitrogen gas. Between the stock solution and pump a gas-trap of a half-filled column (20 cm x 1.5 cm I.D) was inserted. Between pump and injection device, 3 modules were implanted to diminish pump noise mechanically, or hydrodynamically: a coil of 3 meter length (0.25 mm I.D), a Bourdon-pressure gauge, and an unpacked column filled with mobile phase. The nozzle of the polarographic detector was changed by decreasing the outer diameter of the top of the nozzle.

Chemicals

Mitomycin C was kindly supplied by Bristol Myers B.W., Bussum, The Netherlands. Solutions and chemicals were commercially obtained and of analytical grade purity. Mercury was distilled twice before use.

Extraction Procedure

Immediately after blood had been sampled from cancer patients in heparin-containing tubes, the samples were centrifuged and the plasma was transferred to a glass tube and stored at -18°C. For the extraction of MMC 1.0 ml was mixed with 10.0 ml chloroform-2-propanol (1+1, w/w), containing the internal standard. After shaking for 1 min and centrifugation for 5 min (2500 g), the clear supernatant liquid was transferred to a conical glass tube and evaporated to dryness at 30-40°C under nitrogen. The residue was dissolved in 100 µl of methanol. Aliquots of 10 µl were injected into the chromatograph.

RESULTS AND DISCUSSION

The necessity to develop more sensitive and selective detectors in HPLC is widely recognized, especially for the analysis of (non-volatiles) drug metabolites in body fluids. Electrochemical detection with mercury electrodes enables determination of reducible compounds, due to the high overvoltage of hydrogen at mercury. The continuously renewable electrode-surface prevents poisoning of the electrode by reduction products. However, the varying electrode surface gives rise to the same problems as in direct-current polarography: varying faraday currents and charging currents. The SMDE eliminates most of the influences of these processes[14] and is worth optimising as a detector in HPLC. As indicated by most of the cited authors[15-19,21] pump noise and oxygen are the most troublesome parameters. Besides these parameters, the injection system, the nozzle, and some electro-chemical parameters will be discussed.

Oxygen

Due to the reducibility, oxygen interferes in the detection with mercury electrodes. Two types of interference can be noticed. The presence of oxygen in the mobile phase causes a high-background and the presence of oxygen in the sample solution causes a broad, tailing peak in the chromatogram.

Most of the oxygen can be removed from the mobile phase by purging nitrogen through the stock solution, but for analysis at trace level the background is still too high. Hanekamp et al. [24] inserted a scrubber in the flow stream to reduce oxygen, Bond et al. [20] a sulphite bed. Persson and Rosén[19] supplied sulphite ions to the mobile phase and could analyse at microgram level without further elimination procedures. At nanogram level the addition of sulphite ions to the mobile phase only partly removes the oxygen. Deaeration of the mobile phase and the addition of sulphite ions were necessary to get reasonable results. The maximum amount of the sulphite ions was about 0.05 molar, limited by the solubility,

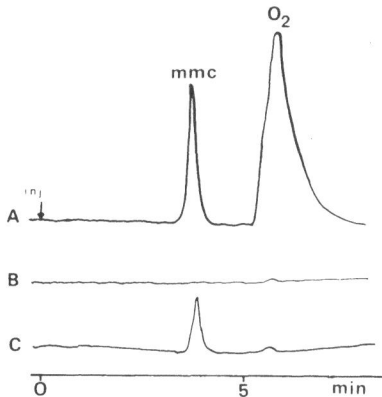

Fig. 1.   A. Chromatogram of 1 µg MMC by detection with SMDE,
          sampled-direct-current mode at -0.6 V. Mobile phase and
          column as indicated in Experimental.   Flow rate 1
          ml/min. No oxygen was removed.   Sensitivity 1 µAf.s.
          B. Blank injection with deaeration of mobile phase and
          sample solution.   Otherwise same conditions as in Fig.
          1.A.
          C. Chromatogram of patient plasma, conditions as in Fig.
          1.B.

depending on the composition of the mobile phase, and limited by the
yield of the oxygen reduction.   The pH-range, in which sulphite can
be used, fits exactly that of MMC: from pH 4 to pH 8.   The applicability of sulphite ions is limited by the reduction potentials of the
compounds to be measured (see Figure 6). Another improvement, related
to sulphite ions, was the shorter stabilizing time of the whole
system, one hour instead of 3-4 hours.

Oxygen proved to be retained in a reversed-phase system, with a
capacity factor of about 1, rather independent of variations of
eluent[19,21] (Figure 1.A.).   Time and amount of sulphite ions was
too small to reduce oxygen from the sample during chromatography.
Deaeration of sample solution was sufficient by use of mobile-phase
saturated nitrogen gas. Because of this it was necessary to introduce
an internal standard during the sample clean up.

The demands on an internal standard are severe: comparable
extraction and chromatographic properties and, especially for use in
the differential pulse mode, similar electrochemical properties.
Only structure-analogs meet these demands.   In this experiment
porfiromycin (PM) was used.

MMC: R-H
PM : R=CH$_3$

The injection system was selected by comparison of a rotary valve injector with a fixed 10 μl injection loop and a Waters U6K septumless injection system. The U6K injection system makes the exclusion of oxygen easier than the rotary valve injector. Care must be taken that the vent of the injector loop is at the same level as the point of injection because of syphonic penetration of oxygen into the sample loop.

It was also necessary to exclude nitrogen or other gases used for deaeration. Local concentrations of micro-gasbubbles at distinct places in the system caused noise (see later). For that reason a gas trap has been included just before the pump in the flow stream. The trap consisted of a large half-filled column with dimensions 20 cm x 1.5 cm I.D.

Pump Noise

Several authors attributed the height of the noise level to pump noise[15-18,21]. Any pump introduces noise into the detection signal. Only a few general rules can be given to select a pump. A single piston (M45, Waters) and a double-piston pump (M6000, Waters) were examined in this study (Figure 2).

The results agreed with those of Persson and Rosén[19]: the double-piston pump caused the least noise, owing to piston-crossover. The noise level was still too high for trace level analysis. The pressure variations were recorded without any damping (Figure 3). Several devices were examined for insertion in the high-pressure part of the system: an empty column, a Bourdonn pressure gauge and a coil of 3 meter length. A packed column was tested too, but was omitted in further studies due to the resultant increase in the total pressure in the system. A combination of an empty column, the pressure gauge and the coil gave the best results (Figure 3). By all these alterations, the current range of 0.2 μA f.s. could be reached in the d.c.-mode, without altering the electronic damping, which would influence the height of the peaks. The relatively low pressure prevented serious interference by local concentrations of small air-bubbles at some places in the system, e.g. in connections and filters. Larger air-bubbles will cause spikes in the signal, due to pressure variations.

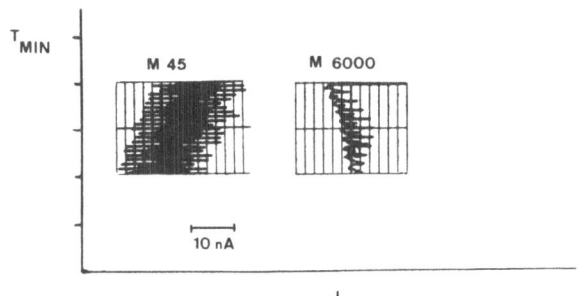

Fig. 2. Recorded detector signals of base line using 2 different
pumps, sampled d.c.-mode, −0.6 V, current range 0.2 µA
f.s., medium drop size, flow 1 ml/min.

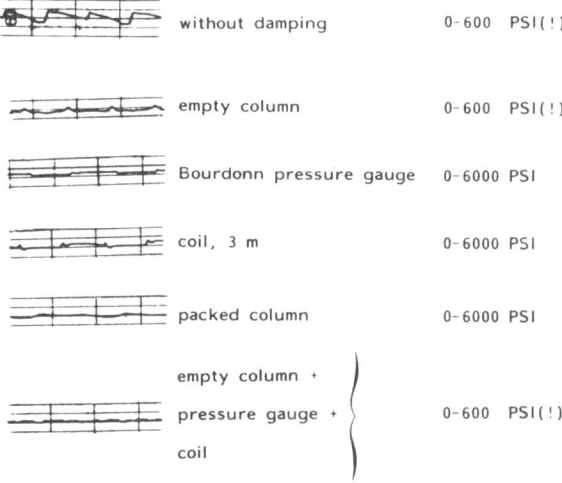

Fig. 3. Pressure signals, with arbitrary units, (at different
PSI-scales), examining several damping devices.

Nozzle

    All nozzles supplied by the manufacturer, gave different
results. All were different in size, inner diameter and resulted in
different flow rates of the eluent, reaching the electrode. One has
to be skilled to place the nozzle reproducibly under the capillary.
When nozzle and capillary are exactly in a straight line, the mercury
drops may fall on top of the nozzle and may cut off the eluent stream
for a longer period. To prevent this, the nozzle has been changed.
Figure 4 shows the supplied and the altered nozzle. Spikes in the
signal can be prevented in this way.

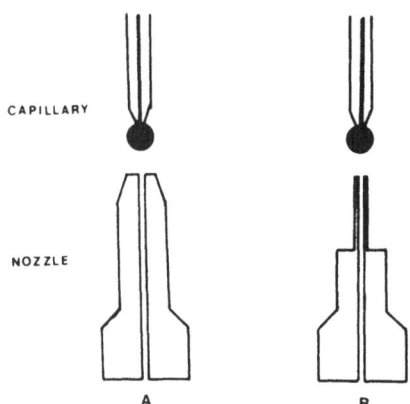

CAPILLARY

NOZZLE

A   B

Fig. 4. The nozzle supplied by the manufacture (A) and the
changed nozzle (B).

Electrochemical Parameters

    The capillary can easily be cleaned and resiliconized due to the
large inner diameter, at least 10 times larger than normal polarographic capillaries. Another improvement was the placing of the
detector into a Faraday cage.

    The potential for the direct-current mode was selected from a
so-called hydrodynamic flow voltammogram (Figure 5). The potential
was chosen on the plateau, before the reduction wave of sulphite ions
(Figure 6). Between pH 7 and pH 9 MMC produces only one reduction
wave[23]. From Figure 5 the pulse base and the pulse height for the
differential pulse mode were selected. The optimum that could be
achieved was -390 mV and -100 mV. To overcome changes in the reference solution of the reference electrode[21], the potential was
checked every day and the inner solution was refreshed every day, if
necessary. Most of the mercury detectors described, make use of very
short drop times, resulting in small electrode-surfaces, fast renewal
of the surface and a low noise level due to low capacity current.
The faraday-current is low too. By using a SMDE, the capacity
current has been practically eliminated at the time of current sampling, and the surface is much larger than normal size capillaries.
Too long drop times (> 3 s) result in a slight decrease of signal due
to poisoning by reaction products or adsorption and by too low
current sampling frequency (the top of the chromatographic peak may
be missed). Enlargement of the drop size leads to higher signals, in
agreement with Vohra and Harrington[18].

    In Figure 7 the relation between flow rate and peak height is
presented. The higher flow rate, the higher signal, due to a thinner
diffusion layer. At flow rates above 2 ml/min the line seems to tend

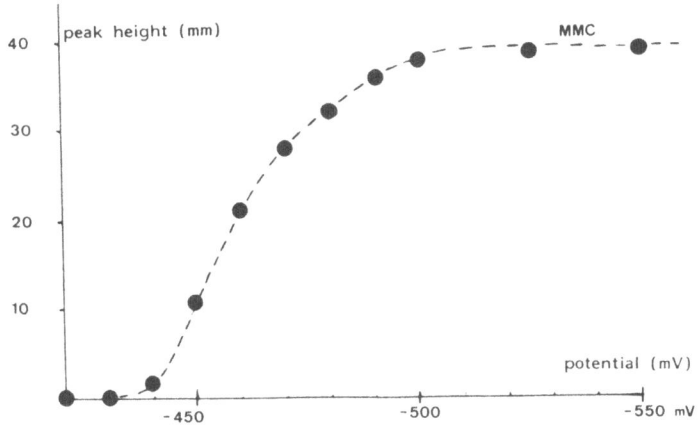

Fig. 5. Hydrodynamic flow-voltammogram of MMC. Injections of
0.50 µg MMC, flow rate 1 ml/min, sensitivity 0.5 µA
f.s., d.c. -mode.

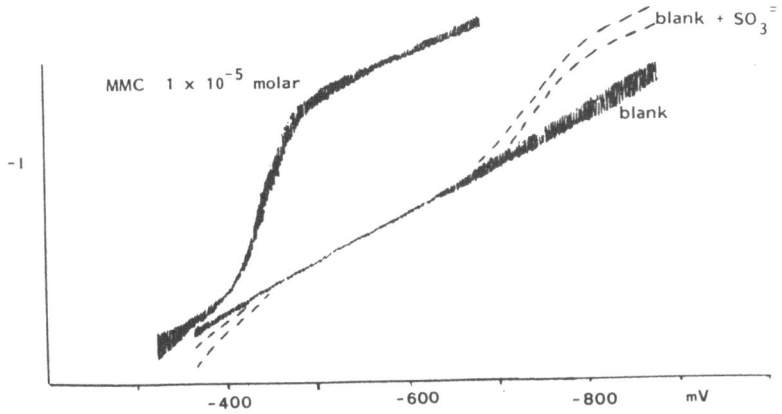

Fig. 6. Polarograms of 10^{-5} molar MMC solutions and blank
solutions at pH 8.0, with and without addition of
sulphite ions.

to a plateau, probably due to too little time being available for
detection. Higher flow rates lead to higher pressure in the system,
and the sensitivity to small air-bubbles is increased, with consequently higher noise levels. The final set-up of chromatographic and
electrochemical parts is given in Figure 8.

Chromatograms and Analytical Data

Figure 9 shows a chromatogram of 500 ng MMC and 500 ng internal
standard, detected in the direct-current mode. Analysis of pure

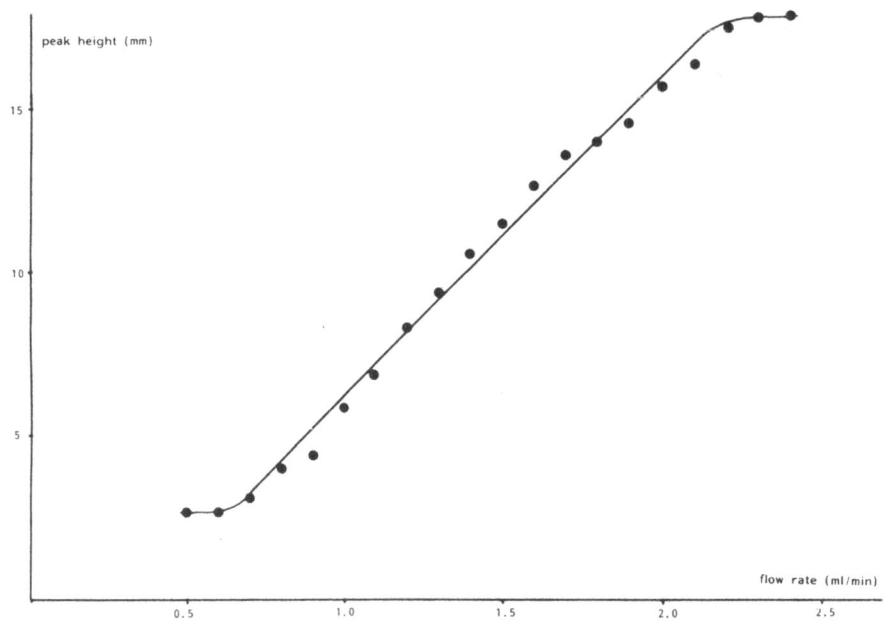

Fig. 7. Relation between flow rate and peak height. Sampled
         d.c.-mode, drop size-large, drop time 1 s, injections of
         0.5 µg MMC.

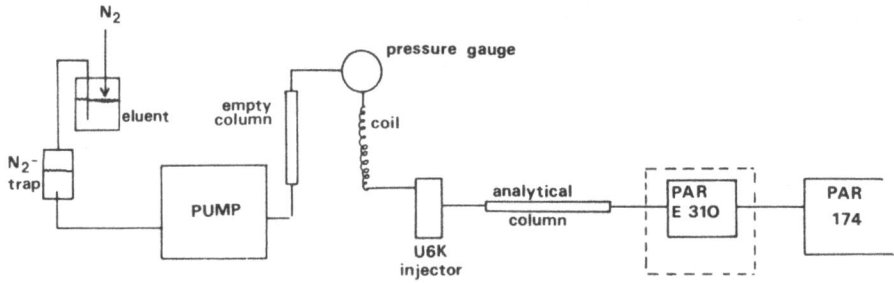

Fig. 8. Total system of chromatographic and electrochemical
         devices.

solutions gave straight lines (r = 0.9993) between 50 ng and 1000 ng
MMC injected. Analysis of serum extracts, according to the procedure, described before[22], gave blank chromatograms, as shown in
Figure 1B. The yield of the extraction procedure is about 90%. The
most sensitive current range was 0.2 µA f.s. Analysis of MMC in the
differential pulse mode gave similar chromatograms; the sensitivity
was enhanced. Straight lines, with comparable reproducibility to
d.c.-detection, were obtained between 5 ng and 1000 ng MMC
(r = 0.9995).

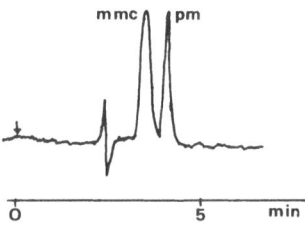

Fig. 9.   Chromatogram of 500 ng MMC and 500 ng porfiromycin,
          detected in the d.c.-mode. Sensitivity 0.5 µA f.s.,
          medium drop size, flow rate 1 ml/min. Mobile phase and
          column as indicated in Experimental.

It can be concluded that optimisation of the damping devices,
the oxygen removal and the nozzle-construction leads to a more reliable detection system and, owing to lower noise levels, to a higher
sensitivity. The selectivity of the detection is so high that the
detection limit of analysis of pure MMC is the same as that of analysis of MMC in plasma extracts, corrected for the yield.

REFERENCES

1. K. Stulík and V. Pacáková, J. Chrom., 208:269 (1981).
2. O. Magnusson, L.B. Nilsson and D. Westerlund, J. Chrom., 221:237
    (1980).
3. K. Brunt, Electrochemical detectors for high-performance liquid
    chromatography and flow-analysis systems, in "Trace
    Analysis", vol. 1, ed. J.F. Lawrence, Academic Press, (1981)
    p. 47-120.
4. S. Ikenoya, K. Abe, T. Tsuda, Y. Yamano, O. Hiroshima, M. Ohmae
    and K. Kawabe, Chem.Pharm.Bull., 27:1237 (1979).
5. K. Shimada, M. Tanaka and T. Nambara, Anal.Lett., 13:1129
    (1980).
6. H. B. Hanekamp, P. Bos, U. A. Th. Brinkman and R. W. Frei, Z.
    Anal.Chem., 297:404 (1979).
7. S. J. Lyle and M. I. Saleh, Talanta, 28:251 (1981).
8. H. B. Hanekamp, W. H. Voogt, P. Bos and R. W. Frei, J.Liq.
    Chrom., 3:1205 (1980).
9. L. Michel and A. Zatka, Anal.Chim.Acta, 105:109 (1979).
10. W. Kutner, J. Debowski and W. Kemula, J.Chrom., 191:47 (1980).
11. P. W. Alexander and M. H. Shah, Talanta, 26:97 (1979).
12. R. Beauchamp, P. Boiney, J.-J. Fombon, J. Tacussel, M. Breant,
    J. Georges, M. Porthault and O. Vittori, J.Chrom., 204:123
    (1981).
13. J. Wang, E. Ouzill, C. Yarnitzky and M. Ariel, Anal.Chim.Acta,
    102:99 (1978).
14. A. M. Bond and R. D. Jones, Anal.Chim.Acta, 121:1 (1980).

15. G. W. Schieffer, J.Chrom., 220:405 (1980).
16. R. Samuelson, J. O'Dea and J. Osteryoung, Anal.Chem., 52:2215 (1980).
17. R. Samuelson and J. Osteryoung, Anal.Chim.Acta, 123:97 (1981).
18. S. K. Vohra and G. W. Harrington, J.Chrom.Sci., 18:379 (1980).
19. B. Persson and L. Rosén, Anal.Chim.Acta, 123:115 (1981).
20. A. M. Bond, H. A. Hudson and P. A. van de Bosch, Anal.Chim.Acta, 127:121 (1981).
21. M. R. Hackman and M. A. Brooks, J.Chrom., 222:179 (1981).
22. J. den Hartigh, W. J. van Oort, M. C. Y. M. Bocken and H. M. Pinedo, Anal.Chim.Acta, 127:47 (1981).
23. J. den Hartigh, R. J. Driebergen, B. Nap, W. J. van Oort and P. Zuman, in preparation.
24. H. B. Hanekamp, W. H. Voogt, P. Bos and R. W. Frei, Anal.Chim. Acta, 118:81 (1980).

ELECTROCHEMICAL DETECTION IN HPLC IN A NATURAL

PRODUCT PROBLEM

Roger M. Smith

Department of Chemistry
University of Technology
Loughborough, Leics., LE11 3TU

Although the emphasis in the application of the electrochemical detector to high performance liquid chromatography[1] has been in the analysis of compounds of biochemical interest such as the catecholamines[2] and estrogens[3], this technique is also applicable to those naturally occurring compounds more commonly described as "natural products". These secondary metabolites are usually less polar than the primary metabolites and are obtained by the extraction of plant material with an organic solvent or steam distillation. Separation of the components then being carried out by crystallisation or chromatography.

Many natural products are commercially important as pharmaceuticals or flavoring agents. The latter include the essential oils which can be readily analyzed by gas liquid chromatography and the pungent principles of the spices, whose determination can often be more difficult because of the involatility of the compounds or their lack of thermal stability. Liquid chromatography would seem to offer an alternative analytical technique and we were interested in its applicability in this field.

Ginger (Zingiber officinale, Zingiberaceae) is grown commercially in may areas of the world and is prepared as either dry powdered ginger, green ginger or crystallised ginger. Two factors contribute to the value of the market spices, firstly the odour produced by the essential oil, which are largely citrals in the higher quality spices, and secondly their pungency [4].

The essential oil can be readily analyzed by gas liquid chromatography [5]. However the pungent principles, which have been identified as a mixture of gingerols (1-111) and related compounds[6,7] are

more difficult to determine as the gingerols readily undergo a
reverse aldol reaction to give zingerone(v) and the corresponding
aldehydes, hexanal-decanal [8] (Scheme 1).   On storage the dehydration of the gingerols yields the corresponding shogaols (IV)   which
have been found to have a weaker pungency [9].

Attempts to carry out gas-liquid chromatography of the gingerols
results in thermal decomposition to zingerone but the reaction is not
reproducible [8].   Some success has been found with the separation of
the gingerols after trimethylsilylation [10].   Thin-layer chromatography enables the groups of essential oils, shogaols and gingerols
to be separated [7,8] but as with many similar series of polar
compounds it is not possible to resolve the individual homologous
gingerols.

This type of separation problem is ideal for reverse-phase
liquid chromatography and a study was undertaken to examine the use
of ODS-Hypersil as a stationary phase.   As the gingerols and shogaols
are phenolic and thus electrochemically active, it was decided to use
ultraviolet and coulometric electrochemical detectors (Kipp model
9205) in series to give absorbance/electrochemical responses.

The extract of ginger was first fractionated by thin-layer
chromatography to give the gingerols.   High performance liquid chromatography using methanol-water (80:20 including 0.01% NaCl) as
eluent gave three electrochemically active peaks when examined using
a potential of 0.8 V (Figure 1).   Preparative separation on hplc and
mass spectrometry confirmed that the compounds were the [6]-, [8]- and
[10]-gingerols.

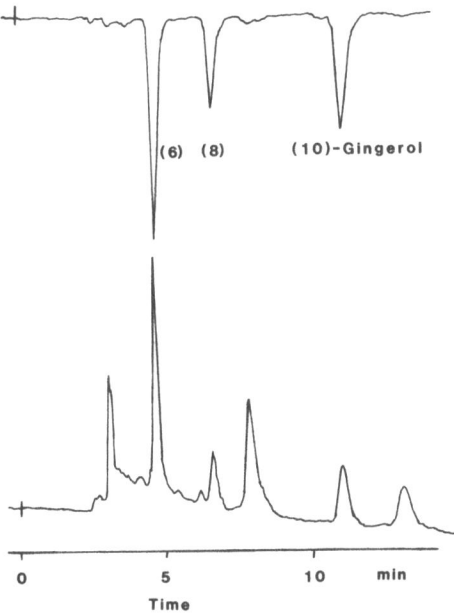

Fig. 1.   Chromatogram of a mixture of [6]-, [8]- and [10]-gingerols.
Column ODS Hypersil, Solvent 80:20 methanol-water (0.01%
NaCl). Flow rate 1 ml min^{-1}. Detector, lower trace 254 nm,
0.08 AUFS; upper trace, Electrochemical 0.8 V 20 µA full
scale.

Comparison with a total extract suggested that a second compound
had the same elution time as [8]-gingerol and this was determined to
be [6]-shogaol by examining other fractions from the TLC separation.
The similarity of the retention of the two compounds was further
checked by comparing their retention indices on the alkylarylketone
scale [11].

In order to obtain a resolution of these two compounds,
acetonitrile-water (60:40 + NaCl) was examined as an eluent system
(Figure 2). The resulting chromatogram showed complete separation of
the four main constituents. The UV/electrochemical ratio confirmed
the identification of shogaol by its higher relative ultraviolet
absorption because of the presence of an unsaturated carbonyl group.

The linearity of the response of the electrochemical detector
was tested using eugenol as a test phenol. On the 20 µA range the
signal was non-linear; however following a 100x dilution a linear
response was obtained.

The method has also been applied to fresh ginger. In this case
the selectivity of the electrochemical detector is particularly

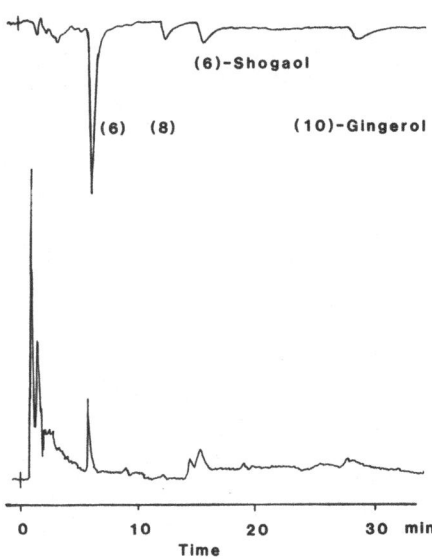

Fig. 2. Chromatogram of a methanol extract of dry powdered ginger,
         column ODS-Hypersil 25 cm x 5 mm. Solvent 60:40
         acetonitrile-water (0.01% NaCl). Flow rate 1 ml/min.
         Detector, lower trace 254 nm 0.01 AUFS; upper trace,
         Electrochemical 0.8 V 2 μA full scale.

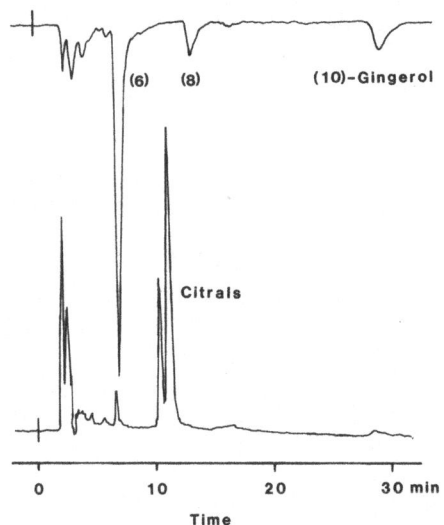

Fig. 3. Chromatogram of methanol extract of fresh ginger. Column
         ODS-Hypersil 25 cm x 5 mm. Solvent 60:40 acetonitrile-water
         (0.01% NaCl). Flow rate 1 ml/min. Detector, lower trace
         254 nm 0.04 AUFS; upper trace, Electrochemical 0.8 V 2 μA
         full scale.

valuable as the UV absorption trace is complicated by major peaks which can be assigned to the citrals (Figure 3). It is notable that the shogaol peak is much smaller than from dried ginger confirming its presence as dehydration product. Other spices including pimenta, and grains of paradise are also being studied.

Acknowledgement

We thank the S.E.R.C. for financial support.

REFERENCES

1. P. T. Kissinger, Anal.Chem., 49:447A (1977).
2. C. J. Little, A. D. Dale and M. B. Evans, J.Chromatogr., 153:381 (1978)
3. P. F. Dixon, P. Lukha and N. R. Scott, Anal.Proc., 16:302 (1979).
4. V.-J. Slazer, Inter.Flavours & Food Add., 206 (1975).
5. R. M. Smith and J. M. Robinson, Phytochemistry, 20:203 (1981).
6. D. W. Connell and M. D. Sutherland, Aust.J.Chem., 22:1033 (1969).
7. D. W. Connell, Flavour Ind., 677 (1970).
8. D. W. Connell and R. McLachlan, J.Chromatogr., 67:29 (1972)
9. P. Denniff, I. Macleod and D. A. Whiting, J.C.S. Perkin I, 82 (1981).
10. D. J. Harvey, J.Chromatogr., 212:75 (1981).
11. R. M. Smith, J.Chromatogr., 236:313 (1982)

ELECTROCHEMICAL SENSORS AND DETECTORS WITH

RENEWABLE ELECTRODE SURFACES

Jiři Tenygl

Czechoslovak Academy of Sciences, The J. Heyrovský
Institute of Physical Chemistry and Electrochemistry
Opletalova 25, 110 00 Prague 1, Czechoslovakia

DISCUSSION OF THE PROBLEM

The passivation of the surface of solid electrodes is one of the most serious problems occurring in the application of electrochemical methods for analysis in practice. The result is a gradual change in electrode activity and corresponding changes of the signal with time. The passivation is often caused by a combination of various factors such as deposition of impurities, formation of insoluble films, oxides, changes in the rate of the electrode reaction by adsorption, deposition of metals changing the chemical composition of the electrode; or by changes of the electrode microstructure and its active area. An increase in the signal is sometimes observed if metals with low overvoltage of hydrogen evolution are deposited or if the electrode surface is increased by corrosion (often observed with carbon electrodes).

The second problem is the slow rate in reaching the steady state signal observed if the potential of the electrode is changed or if the measuring circuit is interrupted. The typical time to reach 95 and 100% steady state values is 5 and 30 minutes, respectively. Shorter, and sometimes even longer, times have been observed, depending on the system and direction of the potential change (often an overshooting of the signal is observed).

That effect does not occur in measurements with dropping mercury electrodes (DME) and a great effort has been employed to develop an electrode with similar features but without the charging current, sensitivity to vibration and to impurities and, of course, without the need to clean the mercury. Such an universally applicable electrode does not yet exist. For some applications, however, electrodes

89

and measuring techniques with similar stability have been developed
and moreover with the possibility to measure at positive potentials.

The method used are mainly based on:

   i   activation of the electrode surface,
  ii   protection of the electrode from passivation,
 iii   combination of several measuring techniques,
  iv   calibration and zeroing of the sensor.

The method of activation will be discussed in more detail later.
The methods under (ii) are applied especially in automatic measurement, industrial sensors and environmental monitors. Those methods
take advantage of e.g. protection of the electrode surface with a
semipermeable membrane, or transfer of the species to be measured
from contaminated solution to the stream of a carrier gas and so on.
The response time of these methods and of automatic titrators and
other system under (iii) is rather long and such methods have not
been included in this paper.

Very good results have been obtained in all system provided with
calibration and zeroing. If the passivation of the electrode is slow
and predictable, it is possible to eliminate its influence on the
accuracy of the measurement by frequent calibration and zeroing. The
introduction of automatic calibration and zeroing and microprocessor
adjustment of the sensor's signal to the correct value represents a
real milestone in the development of electroanalytical methods. It
has been introduced in automatic laboratory measurement, industrial
sensors and environmental monitors, allowing automatic analysis to be
carried out for several weeks or even months with accuracy ± 3%.
Good results have been, however, obtained even with periodical manual
calibration and zeroing.

The methods of automatic EC measurement are scattered in the
literature; useful information can be found in several reviews
[1,2,3,91] or in a prepared monograph[65].

In this paper, sensors and methods suitable especially for
application in chromatographic detectors and fast response sensors
will be discussed.

Electrochemical Detectors for Liquid Chromatography

The basic requirements of detectors for liquid chromatography
and especially HPLC are: fast response, low distortion of the concentration profile of the sample (low washing out of the concentration peaks) and high ratio of analytical to background signal.
These properties are achieved both by proper geometry of the cell and
electrode system and application of a suitable measuring technique.

The volume of the cell must be small (few microliters). These requirements limit the application of methods for maintaining a constant convection and activity of the working electrode. On the other hand the low volume of the cell, without dead space, is convenient for easy calibration and zeroing in a flow-through system. In that case, calibrating and blank solutions are allowed to flow through the detector under the same condition as the sample. The consumption of those solutions is low and they can be introduced periodically between a set of measurements without complicated washing out of the cell or disconnection of the measuring circuit.

Other favorable factors are the low concentration of the species to be measured and the short time of measurement. The passivation is low and for this reason it is often possible to obtain good stability of the measurement even in the case where only occasional activation of the electrode is performed.

The methods of measurement and the various factors which they influence is discussed later on.

Flow-through Voltammetry and Coulometry

The basic electric circuit in flow-through voltammetry and coulometry is the same in principle. The indicating electrode is polarized against a reference electrode to a constant potential. In the close vicinity of the indicating electrode a constant convection is maintained e.g. by a stirrer. The sample flows through the cell and the electric current proportional to the concentration of the species to be measured is recorded. The basic difference between the two methods is in the physical meaning of the current.

In voltammetry, the current is controlled by diffusion of the species to be determined to the electrode surface and is given by the well know equation

$$i = nFDA \frac{C_1 - C}{\delta} \tag{1}$$

where C_1 - concentration of the species to be determined in the sample entering the cell and C_2 in the sample leaving the cell (see further on), C - concentration on the electrode surface (usually C = 0), n - number of electrons, F - Faraday, A - electrode area, δ - thickness of the diffusion layer on the indicating electrode.

The species to be determined is depleted by electrolysis and the limiting current i_ℓ in time t is given by

$$i_{\ell,t} = i_{\ell,t=0} \exp - \frac{DA}{V\delta} t \tag{2}$$

The equation (2) express the exponential decrease of the concentration with time given roughly by V/v, where V is the volume of the cell and v the flow rate.

Voltammetric sensors are constructed in such a way that very little depletion of the species to be determined occurs and C_1 is nearly equal to C_2. That condition is maintained in most chromatographic detectors, where A and V are small, and v high. The δ is inversely proportional to the convection and must be constant, as well as to the activity of the working electrode.

A typical feature of voltammetric sensors is fast response, linear response in a broad concentration range, non-linear dependence on the flow rate and the temperature dependence of the signal (+2 - 5%°C).

Coulometric sensors are constructed, on the other hand, so that practically all the species to be determined is depleted by electrolysis during passage through the sensor and $C_2 < 0,01\ C_1$. At the constant v the current is given by Faraday law

$$I = nFN = nFC_1v$$

where N is the number of moles of the species to be determined entering the cell in a time unit. In the coulometric mode of operation, the current is controlled by the feed rate of the sample and corresponds to the charge Q passed in a time unit $Q = it$. The important feature of the coulometric sensor is that the current is, within certain limits, independent of the change of the working electrode activity. The accuracy in sample flow rate controls the accuracy of measurement and must be kept as constant as possible; the signal is, however, independent of the temperature. Convection must be kept constant to prevent corresponding fluctuations in the signal. That fact is important in the application to chromatographic detection as fluctuations of the signal due to the change in convection cannot be filtered out by e.g. damping of the recorder without reducing the response rate.

There is a dilemma: in flow-through voltammetry is necessary to maintain a constant activity of the working electrode, in flow-through coulometry the sample flow rate must be kept constant.

To reach the state of complete or nearly complete depletion of the species to be determined without undue increase in the response time in flow-through coulometry is a difficult task. It is necessary to prevent mixing of the sample entering the cell with the sample leaving the cell and to maintain a high and reproducible convection. That requirement is often achieved by prolonged contact of the sample with the working electrode surface using a labyrinthine path of the solution or by using a system of tubular or porous electrodes. In that case, an undesirable increase of the residual current due to the large working electrode surface is often observed

Both methods offer advantages and limitations and, as usual, it is necessary to choose the appropriate methods for a particular application. Flow-through coulometry found use especially in the long-term analyzers for industry and environmental monitoring while flow-through voltammetry mainly in detectors and sensors for automated analysis.

That classification is only a rough guide to the application of EC methods in practice. Many other problems are involved. One problem is that the classification of the sensors in the literature is not uniform. The term "coulometric" is usually used if an electrolytic generation of the titrant is employed with potentiometric or amperometric detection of the point of equivalence, which is, in fact, an automatic flow-through titrator. The same is valid for the mixed use of the terms "polarographic" and "voltammetric" which should use for sensors with DME and solid electrodes, respectively. Another problem is that many sensors and commercial detectors work somewhere in between voltammetry and coulometry and the species to be determined is depleted by the passage through the detector by the factor 0.2 - 0.6. Often, the efficiency of the depletion varies with concentration and flow rate; these detectors are difficult not only to classify but to use in practice.

Factors Governing the Stability

The basic condition for stability is that the products of the electrode reaction or intermediates are soluble and do not form a passivating film. Another important factor governing the stability is the potential of the electrode. Little or no passivation at all can be expected in the potential region + 0.1V to + 0.7V SCE. The electrode surface is not corroded by anodic oxidation (causing a gradual increase of background current), the background current itself is low and the deposition of traces of heavy metal on the electrode surface does not take place. Many organic substances are oxidized in that potential region without interference from atmospheric oxygen which need not be purged from the solution. In the same potential region oxidation of halide (I^-, Br^-) takes place or reduction of halogen (I_2, Br_2) if preliminary chemical reaction is used. Inert electrodes such as carbon (pyrolytic carbon, glassy carbon, carbon paste), platinum and gold are used.

In the negative potential region (0.0V to -1.8V SCE) passivation is often observed and the electrode surface has to be regenerated periodically or, preferably, continuously. Accumulation of metals (especially Cu, Ni, Fe) on the electrode surface causes a decrease of the overvoltage of the hydrogen evolution and a gradual increase in

the background current. Further deformation of the horizontal
plateau of the limiting current is observed. The electrode material
is usually mercury with the highest hydrogen overvoltage and most
negative working potential limit (about -1.8V for DME and -1.0V for
mercury pool electrode depending on pH).

Contribution of Convection

Intensive and reproducible convection of the sample in the close
vicinity of the working electrode surface is necessary for obtaining
a fast response and a reproducible signal. The field of hydrodynamic
voltammetry[21] with rotating disc electrode[22] is a typical representative offering the possibility of mathematical description of the
convection. In practice, other methods such as vibrating electrodes,
stirring of the solution with stirrers of various shapes[23-27], fast
flow of the electrolyte through the cell[28-33], circulation of the
solution with a pump[34], bubbling of the stream of a gas through the
sample[35] and electrodes of various shapes (wire, spherical, conical) have all been used. Those methods and electrodes have been used
to simplify the construction and eliminate the trouble with sliding
contact and are widely used in practice. The mathematical description was often not given but the convection is reproducible. In the
early papers, some authors tried to use the fast movement of the
electrode or solution for simultaneous activation of the electrode
surface. It should be mentioned here that that sort of activation is
ineffective.

In chromatographic detectors with very low volume only a few
methods can be used. Convection is often maintained by the flow of
the sample at a constant rate through the cell. A tubular[36-38,41]
or wall-jet electrode[13-15,31-33,39,40,42-45,92,93] used. The
tubular electrode made of a seamless platinum tubing or drilled
carbon rod (i.e. about 1 mm), makes it possible to determine the
background current of the sample by the stopped-flow technique[37,38]
but it is difficult to activate the inner wall of the tube serving as
the electrode surface. Favorable current distribution, easy mechanical cleaning, fast response and low peak distortion are offered by
the wall-jet electrode used in a chromatographic detector[5]. Detectors of various construction (with thin layer of electrolyte flowing
through) have been described[46-52]. The signal is dependent on the
flow rate which should be kept constant. To increase sensitivity and
reduce the flow rate dependance of the signal, measurement in pul-
sating flow was introduced[6]. The effluent leaving the column is
pumped with a simple peristaltic pump forward and backward at a
frequency of 5 - 20 cycle/sec while the positive forward rate is
maintained.

The increase of the limiting current of the DME depends on the
direction of the flow, drop time, shape of the capillary orifice and

cell design and is in laminar flow about 20-30% for a flow rate of
0.50 cm/sec. At higher flow velocity, mercury drops are torn off the
capillary and irregular dropping is observed causing fluctuation of
current. To prevent that effect, the capillary is usually shielded
or situated in a by-pass. Those methods are used in automatic long
term analysis. In chromatographic detectors, a horizontal Smoler's
capillary is often used. Turbulence of the solution is suppressed by
proper design of the cell. Measurement at a velocity up to 10 cm/sec
is possible. Drop stability can be also increased by mechanical
control of the drop time causing a prematurely dislodged drop.

The requirement for low cell volume complicates the design of
detectors. Another approach to solve that problem is measurement in
a foam[9]. The effluent is mixed with a detergent and a stream of
nitrogen. A large volume foam is formed and led to the voltammetric
or potentiometric detector. The soap bubbles travel along the electrode system and regenerate the thin film of the electrolyte on the
electrode surface. Due to the high (5000:1) gas to liquid mixing
ratio, fast response is obtained even in a detector with large volume.

Dropping Mercury Electrode

The DME offers the most negative working potential range, regenerated surface and reproducible convection caused by the growth of
the drop. The measurement of low concentrations (10^{-5} - 5.10^{-7}M) is
hampered by the charging current and by the so-called "capillary
noise". It was shown that both factors can be suppressed by proper
capillary design and novel types of capillaries have been introduced.

A recent innovation is the static mercury drop electrode [55].
The mercury flow to the capillary is controlled by an electromagnetic
valve opened periodically for a short time only so that a mercury
drop of a constant area is formed. The current is registered some
time after the drop formation when the charging current is reduced
practically to zero and does not interfere. The measurement is
performed by d.c. or differential pulse technique and the sensitivity
of the measurement is highly increased. A chromatographic detector
with SMDE is manufactured commercially [54].

Another approach to the compensation of the charging current has
been used in the pump-fed DME [7]. Mercury is fed to the capillary
with a simple but precise peristaltic pump and a mercury drop of a
constant size is formed as in SMDE. After some time when the charging current decays, the pump forces mercury forward (t_1) and backward
(t_2) so that the size of the static drop is increased and decreased
back to its original size. The charging current flows in a positive
direction when the drop increases and negative direction when decreased and is thus cancelled. By integration between t_1 and t_2 only

the Faradaic charge of analytical interest is obtained. The difference from the SMDE is that the measurement is made on a moving mercury surface so that the depletion of the species to be measured around the drop is eliminated and the sensitivity of the measurement increased but without interference from the charging current. A pulsed-flow DME was used in the study of the adsorption process[16] and an interrupted flow DME has been studied for general use [19].

Novel constructions of the DME capillary have been introduced as well with the aim of increasing the stability of the drop and reducing the capillary noise. The problem is to eliminate penetration of the solution into the capillary between the mercury and the glass causing fluctuation of the current. That effect can be observed especially at potentials -1.4V and more negative and is believed to be caused by several factors. The most important of these were cracks in the glass at the capillary orifice [55,56], the cylindrical shape of the capillary[57] made from thermometer capillary and still widely used to the present day instead of the hand drawn conical capillaries used by Professor Heyrovský at the beginning of polarography[58]. A remarkable increase in stability was obtained by making a spindle-shape cavity at the capillary orifice [11]. Of importance is the conical shape of the mercury thread in the capillary near its orifice. Hydrophobization of the capillary has a limited effect only and the results obtained are controversial [56, 59,60]. Capillaries of polyethylene[20], plastic[18], glass covered with polyethylene[17,61] and an easy-to-make capillary using the Technicon Analyzer tubing have been recently introduced for measurement in glass-corrosive media. Automated analysis with DME has been reviewed [62].

Activation of the Electrode

Activation of the electrode has been studied since the very beginning of electro-analysis and various methods have been introduced. We will discuss in brief mainly the principles applicable for laboratory measurement and especially for chromatographic detectors. There is a variety of other methods and discussions of their application in practice is scattered in the literature[1-3,65], but a comprehensive review has appeared recently [60,63].

Mechanical Methods

Mechanical cleaning by polishing with abrasive material ranks among the oldest methods and have been used by many authors up to the present. It can be easily done manually while an automatic mechanical cleaning is quite a complicated procedure. The reason is that the electrode and the cleaning tool must not be in constant contact but move independently against each other to prevent the deposition of the abraded electrode material on the tool. The accumulated abraded material adhering to the tool is in electrical contact with the electrode and increases its area irreproducibly.

A simple method fulfilling the requirements seems to be the rotation of the Pt wire electrode in a suspension of carborundum powder [64]. It is cleaned continuously by the bombardment of solid particles. [For other methods see 63.]

The efficiency and success of mechanical polishing with abrading paste or very fine emery paper depends on the skill of the personnel: abrading must not either change the microstructure of the electrode surface nor its wetting angle by, e.g., touch of fingers. The controversial results in the literature support the fact that the electrodes of commercially available sensors are sometimes damaged after a certain time of use.

Electrochemical Methods

Activation of the working electrode is performed by the electrode being periodically disconnected from the measuring system and polarized by one or more cathodic and/or anodic pulses.

The inconclusive results obtained with that technique need to be discussed.

The efficiency of EC methods for electrode activation depends mainly on[65]; i) electrode material, ii) solution composition (pH, complexing agent), iii) potential of the cleaning pulse, iv) convection of the solution in the close vicinity of the electrode.

The obvious and widely accepted mechanism is the anodic dissolution of cathodically deposited films and metals. That process occurs, however, only in a limited number of application when the composition of the solution makes possible a reversible anodic reaction. In other cases, a not generally known trick consisting of the addition of a complexing agent to the solution, can help. The complexing agent prevents first the deposition of metal by shifting the deposition potential of metal ions to more negative value and (or) enables the anodic dissolution. NH_4^+/NH_4OH buffer or strong complexing agent (EDTA) has a beneficial effect in preventing the deposition of Cu and other metals with low hydrogen overvoltage. Of importance is the potential of the cleaning pulse as well. If oxygen

$$2H_2O = 4H^+ + O_2 + 4e$$

or hydrogen

$$2H_2O = H_2 + 2OH^- + 2e$$

evolution takes place, the limited mechanical cleaning by action of
the gas bubbles is combined with consequent acidification or alkaliz-
ation of solution in the close vicinity of the electrode surface.
The local change of pH is obtained only if the convection is stopped
and can have either beneficial (dissolution of films) or detrimental
effect (precipitation of hydroxides; precipitation of reaction intermediates) depending on the composition of the solution. Polarization
to very positive potentials causes corrosion of the electrode itself
(formation of oxide films, disintegration of carbon) and should be
avoided. It seems, that the best results are obtained with automatically applied potentiostatic pulses and a measuring period taking
place at a defined time after the pulse application. Electrochemical
activation is usually combined with a periodical mechanical cleaning.

With a mercury pool electrode (MPE), good results have been
obtained by adding about 1.10^{-4}M Hg^2 to the supporting electrolyte[74,75]. The fresh layer of mercury continuously deposited on
the MPE covers the deposited impurities and prevents the deformation
of the limiting current plateau.

Thermic Regeneration

Cleaning of Pt electrodes by annealing in the flame of a Bunsen
burner is one of the oldest methods. The results obtained are,
however, rather irreproducible. Heating of a Pt electrode made of
wire in the form of a loop by passage of an electric current was used
to pyrolyze the passivating film[68] and in an attempt to cause the
thermic convection of the solution[69]. A remarkable improvement was
obtained in the method of automatic thermic regeneration [8,70,71].
The Pt electrode in form of a loop is mechanically lifted above the
solution level, heated to a red glow by passage of an electric current, brought back to the solution, pretreated by current pulses and
then used for EC measurement. That procedure is repeated periodically every 2-6 sec.

The measurement itself is made at a defined time after the
heating of the electrode to a defined temperature and at a defined
time after polarization of the electrode. This results in a remarkable increase in the measurement reproducibility. The heated wire
electrode (HWE) behaves as an electrode without history so it is even
possible to make a differential measurement at several different
potentials. The reactivation of the electrode depends on the temperature and composition of the gas. All organic substances including
the passivating phenolic film are pyrolyzed by heating to about 600°C
(red glow). A new effect is observed if the electrode is heated to
about 1100°C in air and platinum oxide is reduced by a cathodic pulse
before measurement. The electrode becomes catalytically active and
its activity is regenerated in every cycle. A number of species
being electrochemically inactive on the ordinary electrode can be

then determined voltammetrically or potentiometrically (alcohols, CO, H_2 etc).

Besides the described mechanical lifting of the electrode, hydraulic and pneumatic systems have been further employed and a chromatographic detector constructed [72].

Mercury Electrodes

The typical example is the mercury pool electrode[73] (MPE), introduced for higher sensitivity over the DME because of its larger surface and very low background current and, of course, to eliminate the cleaning of mercury. MPE of various designs (layer of Hg on the bottom of a cylindrical cell, mercury level in wide bore vertical tubing, mercury drop in a semispherical housing with Pt contact) and size (diameter 8-20 mm for analytical application in stirred solution[74,75], up to 40 mm in long term measurement[76,77], 0.5-5 mm in chromatographic detector)[78]. There is no sharp dividing line between MPE and stationary mercury drop in a large bore capillary pointing upward.

The intrinsic disadvantage of the stationary mercury electrode is that the working surface of the electrode is the mercury level. Impurities of lower specific gravity, metals or precipitates float on the surface and interfere with measurement. It is difficult to remove them due to the high surface tension.

In view of this neglected phenomenon it is possible to predict the efficiency of the activating methods. Mechanical stirring of the mercury pool has only a limited effect. The same is valid for methods not involving regeneration of the surface layer as in addition of fresh mercury to the mercury pool[79], slow flow of mercury under the meniscus[80], flow of mercury through a tube with a small hole in the wall[81]. The efficient method is the overflow of mercury from the meniscus[72,77] corresponding, in effect, to the replacement of a stationary drop. The consumption of mercury is however quite high. An effective method is also the cathodic deposition of mercury as previously discussed [66,67].

Another principle that has been used is the rotating mercury electrode with regenerated surface[12,83]. The working surface is the bottom of mercury layer separated from the sample by a cellophane membrane. Amalgams or other impurities are transferred to the mercury surface where they are continuously dissolved in a layer of a cleaning solution.

Solid electrodes covered with a layer of mercury have been used especially for making large surface electrodes of any geometry [84-86]. Those electrodes are used especially in anodic stripping voltammetry[87-90].

These new achievements suggest that further developments of EC methods can be expected. The author believes that a fully automatic system with recirculation of mercury in a completely closed system will be soon introduced.

Acknowledgement

The author would like to thank the organizing committee of the 5th Anglo-Czech meeting and especially to Dr. T. H. Ryan for all the effort and help in publishing of this paper.

REFERENCES

1. W. J. Blaedel and R. H. Laesing, Automation of the analytical process through continuous analysis, in: "Advances in Anal. Chem. and Instrumentation," Vol.5, p.66, C. N. Reilley and F. W. McLafferty, eds., Interscience Publ., New York, (1966).
2. P. Hersch, Galvanic analysis in advances in Anal. Chem. and Instrumentation, Vol.3, p.183, J. Wiley & Sons, New York, (1964).
3. J. T. Stock, Amperometric Titration, Interscience Publ., New York, (1975).
4. B. Fleet and C. J. Little, J.Chromat.Sci., 12:747 (1974).
5. EDT Research Inc., 14 Trading Estate, London N.W. 107LU.
6. J. Tenygl, J. Heyrovský Memorial Congress on Polarography, Prague, Proceedings 2, p.173, Aug.25-29, (1980).
6a. J. Tenygl, J. Heyrovský Memorial Congress on Polarography, Prague, Proceedings 1, p.55, Aug.25-29, (1980).
7. J. Tenygl and B. Fleet, J. Heyrovský Memorial Congress on Polarography, Prague, Proceedings 2, p.155, Aug.25-29, (1980).
8. J. Tenygl, J. Heyrovský Memorial Congress on Polarography, Prague, Proceedings 2, p.172, Aug.25-29, (1980).
9. J. Tenygl, J. Heyrovský Memorial Congress on Polarography, Prague, Proceedings 2, p.171, Aug.25-29, (1980).
10. J. Tenygl, J. Heyrovský Memorial Congress on Polarography, Prague, Proceedings 2, p.172, Aug.25-29, (1980).
11. L. Novotný, J. Heyrovský Memorial Congress on Polarography, Prague, Proceedings 2, p.129, Aug.25-29, (1980).
12. J. Tenygl, J. Heyrovský Memorial Congress on Polarography, Prague, Proceedings 2, p.171, Aug.25-29, (1980).
13. A. Ivaska and T. H. Ryan, J. Heyrovský Memorial Congress on Polarography, Prague, Proceedings 2, p.72, Aug.25-29, (1980).
14. M. Breant, J. Georges, O. Vittorio, J. J. Fombon, and J. Tacusel, J. Heyrovský Memorial Congress on Polarography, Prague, Proceedings 2, p.22, Aug.25-29, (1980).

15. R. Beauchamp, P. Boinay, J. J. Fombon, and J. Tacusel, J.
     Heyrovský Memorial Congress on Polarography, Prague, Pro-
     ceedings 2, p.12, Aug.25-29, (1980).
16. C. Buess-Herman and L. Gierst, J. Heyrovský Memorial Congress
     on Polarography, Prague, Proceedings 2, p.25, Aug.25-29,
     (1980).
17. H. Menard and B. Dubrenil, J. Heyrovský Memorial Congress on
     Polarography, Prague, Proceedings 2, p.116, Aug.25-29,
     (1980).
18. L. Novotný, J. Heyrovský Memorial Congress on Polarography,
     Prague, Proceedings 2, p.130, Aug.25-29, (1980).
19. L. Novotný, J. Heyrovský Memorial Congress on Polarography,
     Prague, Proceedings 2, p.129, Aug.25-29, (1980).
20. L. Novotný, J. Kuta, and I. Smoler, J.Electroanal.Chem., 88:161
     (1978).
21. V. C. Lewich, Physicochemical Hydrodynamics, Prentice Hall, New
     Jersey, (1962).
22. J. Newman, in: "Electroanalytical Chemistry", J. A. Bard, ed.,
     Vol.6, p.187, Marcel Dekker, New York, (1973).
23. D. J. Rosie and W. D. Cooke, Anal.Chem., 27:1360 (1955).
24. I. M. Kolthoff, J. Jordan, and S. Prager, J.Am.Chem.Soc.,
     76:5221 (1954).
25. A. D. Adamse, J.Water Pollut.Control.Fed., 37:1481 (1965).
26. P. Arthur, I. C. Komyathy, R. F. Maness, and H. W. Vaughan,
     Anal.Chem., 27:895 (1955).
27. J. Svancer, J. Kupec, and L. Kupcová, Kozarství (Czechoslovakia)
     20:191 (1970).
28. J. Jordan, R. A. Javick, and W. E. Ranz, J.Am.Chem.Soc., 80:3846
     (1956).
29. T. O. Oesterling and C. L. Olson, Anal.Chem., 39:1543,1546
     (1967).
30. W. J. Blaedel and S. L. Boyers, Anal.Chem., 43:1538 (1971).
31. B. Fleet and N. B. Fouzder, J.Electroanal.Chem., 63:59 (1975).
32. B. Fleet and N. B. Fouzder, J.Electroanal.Chem., 63:69 (1975).
33. B. Fleet and N. B. Fouzder, J.Electroanal.Chem., 63:79 (1975).
34. V. I. Kuleshov and V. M. Pichugina, Zavod.Lab., 34:1557 (1968).
35. J. V. A. Novák, Chem.listy, 49:1476 (1955).
36. W. J. Blaedel and S. L. Boyers, Anal.Chem., 43:1538 (1971).
37. W. J. Blaedel, C. L. Olson, and L. R. Sharma, Anal.Chem.,
     35:2100 (1966).
38. W. J. Blaedel and S. L. Boyer, Anal.Chem., 43:1538 (1971).
39. J. Yamada and H. Matsuda, J.Electroanal.Chem., 44:189 (1973).
40. B. Fleet and C. J. Little, J.Chromatographic Sci., 12:747
     (1974).
41. K. Stulík and V. Hora, Electroanal.Chem., 70:253 (1976).
42. H. Matsuda, J.Electroanal.Chem., 15:109 (1967).
43. A. J. Dikusar and M. B. Bardin, Zhur.Anal.Khim., 26:1059,1068
     (1971).
44. H. Matsuda and J. Yamada, J.Electroanal.Chem., 30:261,271
     (1971).

45. M. Varadi and E. Pungor, Anal.Chim.Acta, 94:351 (1977).
46. D. G. Schwartzfager, Anal.Chem., 48:2189 (1976).
47. P. L. Joynes and R. L. Maggs, J.Chromatogr.Sci., 8:427 (1970).
48. P. T. Kissinger, L. J. Felice, R. M. Roggin, L. A. Pachla, and
    D. C. Wenke, Clin.Chem., 20:992 (1974).
49. S. E. Magic, J.Chromatogr., 129:73 (1976).
50. Sasa Suleiman and C. L. Blank, Anal.Chem., 49:354 (1977).
51. C. L. Blank, J.Chromatogr., 117:35 (1976).
52. Bionalytical Systems, Inc., Lafayette, Ind., U.S.A.
53. Princeton Applied Research Ltd., Princeton, N.J., U.S.A.
54. Princeton Applied Research, Model 310 Polarographic Detector.
55. I. Smoler, private communication.
56. T. A. Kryukova, G. T. Sinyakova, and T. V. Arefyeva,
    Polyarograficheskiy analiz., Gos.Nauch.-tekhn.Izd.Khim.Lit.,
    Moskwa, p.128 (1959).
57. H. Siebert and T. Langer, Chem.Fabrik., 11:141 (1938).
58. J. Heyrovský, Polarographie, Springer, Wien, (1941).
59. G. C. Baker, Anal.Chim.Acta, 18:118 (1958).
60. W. D. Cooke, M. T. Kelley, and D. J. Fisher, Anal.Chem., 33:1209
    (1961).
61. H. Menard and F. Leblond-Routhier, Anal.Chem., 50:687 (1978).
62. J. Tenygl, Talanta, in press (1982).
63. J. Tenygl, Zavod. Lab., in press (1982).
64. J. Tenygl, These kandidátské disertacní práce, Cesk.Akad., Ved,
    Praha, (1961).
65. J. Tenygl and B. Fleet, Automatic Electrochemical Analysis,
    M. Dekker, New York, in press.
66. J. Tenygl, Czech. Patent No.106 185, Appl. April 14, (1961).
67. J. Tenygl, Coll.Czech.Chem.Comm., 37:701 (1972).
68. V. E. Shevtsov, Zavod.Lab., 37:785 (1971).
69. L. Ducret, Compt.Rend., 252:1948 (1961).
70. J. Tenygl and J. Vána, Czech. Pat. Appl. PV 7588 (1977); Czech.
    Pat. Appl. PV 3629 (1978).
71. J. Vána and J. Tenygl, Czech. Pat. Appl. PV 2022 (1979).
72. J. Tenygl, not yet published.
73. C. A. Streuli and W. D. Cooke, Anal.Chem., 25:1691 (1953).
74. C. A. Streuli and W. D. Cooke, Anal.Chem., 26:963 (1954).
75. D. J. Rosie and W. D. Cooke, Anal.Chem., 27:1360 (1955).
76. J. V. A. Novák, Chem.listy, 49:1476 (1955).
77. J. Závorka and F. Stráfelda, Chem.listy, 51:2374 (1957).
78. D. L. Rabenstein and R. Saetre, Anal.Chem., 49:1036 (1977).
79. M. P. Simonin, J.Chim.Phys., 57:161 (1960).
80. V. S. Griffiths and W. J. Parker, Anal.Chim.Acta, 14:194 (1956).
81. D. J. Ferret and C. S. C. Phillips, Trans.Faraday Soc., 51:980
    (1955).
82. G. E. Penketh, J.Appl.Chem.(London), 10:324 (1960).
83. J. Tenygl, Czech. Patent 180 448 (1976).
84. E. M. Skobets, L. S. Beremblyum, and N. N. Atamanenko,
    Zavod.Lab., 14:131 (1948).
85. T. L. Marple and L. B. Rogers, Anal.Chem., 25:1351 (1953).

86. E. D. Cooke, Anal.Chem., 25:215 (1953).

87. V. A. Igolinskii and A. G. Stromberg, Zavod.Lab., 30:656 (1964).

88. V. G. Nagaev, M. S. Zakharov, and V. V. Pnev, Zh.Anal.Khim., 25:1450 (1970).

89. T. M. Florence, J.Electroanal.Chem.Interfacial Electrochem., 26:293 (1970).

90. A. M. Hartley, A. G. Hiebert and J. A. Cox, J.Electroanal.Chem. Interfacial Electrochem., 17:81 (1968).

91. R. N. Adams, Electrochemistry at Solid Electrodes, M. Dekker, New York (1969).

92. T. H. Ryan and P. S. Wilson, Laboratory Practice, 501 (May 1969).

93. B. Fleet, H. Gunasingham, G. G. de Gamia, T. A. Berger, S. das Gupta, and C. J. Little, in press.

VOLTAMMETRY OF ORGANIC MOLECULES AT SOLID ELECTRODES

J. Volke

J. Heyrovský Institute of Physical Chemistry and
Electrochemistry, Czechoslovak Academy of Sciences
118 40 Praha 1

INTRODUCTION

The application of electrochemical detectors, e.g. in combination with chromatography, in case of new, as yet uninvestigated molecules, is often rendered difficult because of troubles due to predicting the electroactivity of the substance to be investigated or to be determined. This is particularly true with solid, non-mercury electrodes. For this reason most problems of this type are solved empirically and the firms producing analytical instrumentation (such as HPLC with electrochemical detectors) offer their customers the performance of preliminary tests in order to detect activity at the particular electrode material. For constructional reasons solid electrodes are usually preferred to mercury electrodes (i.e. to the dropping mercury electrode). This represents of course a serious drawback since the prediction of reducibility or oxidisability at a mercury electrode is based on a vast quantity of experimental data obtained chiefly from polarographic measurements[1]. The quoted tables[1] contain also to a lesser degree values obtained with platinum and similar electrodes. When working with mercury electrodes, we know all the active (mostly reducible) groups, the influence of the molecular framework on the behavior of such a group, the mechanism of the electrode reaction and the relationship between the polarographic behavior and structure. The electroactive groups and the corresponding mechanisms on mercury, resulting from preparative electrolyses and from polarography have been often tabulated (cf.[2]) Table 1.

The situation is somewhat different with solid electrodes: several materials are used here such as platinum, gold, as well as different types of carbon materials, e.g., glassy (vitreous) carbon etc. All these solid electrodes are particularly convenient in

Table 1.   Organic Electroactive Groups on Mercury Electrodes

Bond Grouping	Example
C – N	$R-CH_2-Br \xrightarrow{2e^-,H^+} R-CH_3 + Br^-$ (structures)

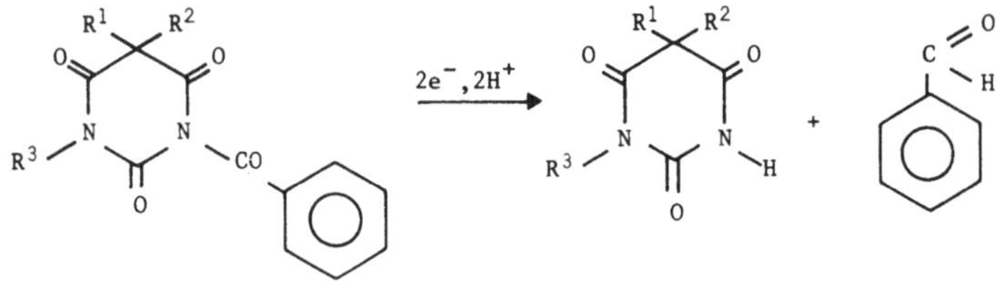

C – N

$$\xrightarrow{2e^-,2H^+}$$

C – O

$$\xrightarrow{2e^-,H^+}$$

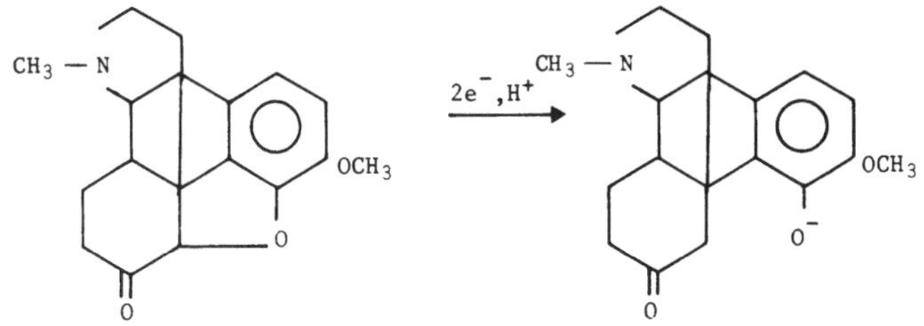

C – Hal

$$R-CH_2-Br \xrightarrow{2e^-,H^+} R-CH_3 + Br^-$$

$$C_6H_6Cl_6 \xrightarrow{6e^-} C_6H_6 + 6Cl^-$$

N – N

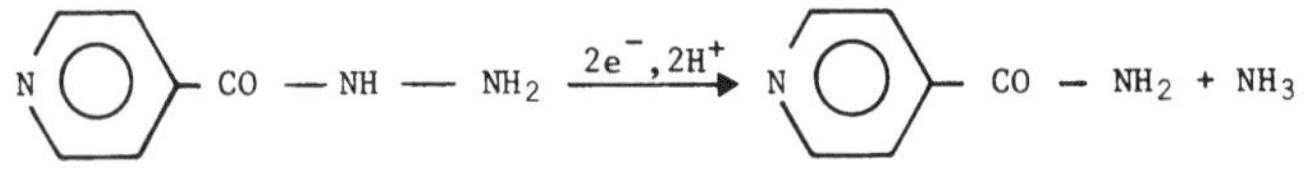

$$\text{N} \bigcirc -CO-NH-NH_2 \xrightarrow{2e^-,2H^+} \text{N} \bigcirc -CO-NH_2 + NH_3$$

O – O

$$O-O \xrightarrow{2e^-,2H^+} HO \quad OH$$

S – S

$$HOOC-\underset{NH_2}{CH}-CH_2-S-S-CH_2-\underset{NH_2}{CH}-COOH \xrightarrow{2e^-,2H^+}$$

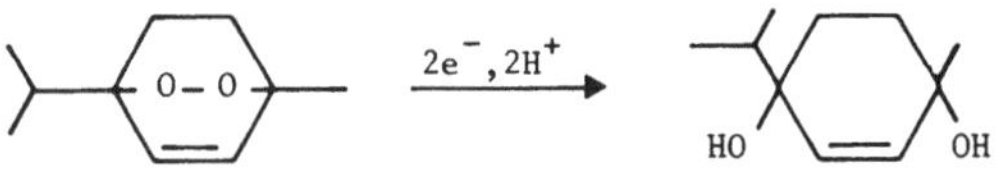

$$2HS-CH_2-\underset{NH_2}{CH}-COOH$$

C = C

$$RO-\bigcirc-CH=CH-\bigcirc-OR \xrightarrow{2e^-,2H^+}$$

$$RO-\bigcirc-CH_2-CH_2-\bigcirc-OR$$

C = O

$$F-\bigcirc-CO-CH_2-CH_2-CH_2-N\bigcirc N-\bigcirc-OCH_3$$

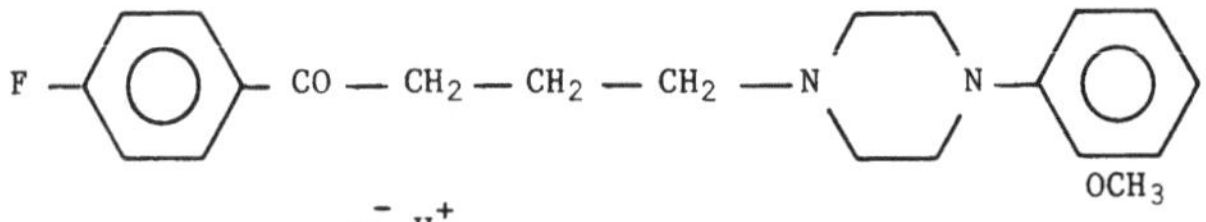

$$\equiv Ar-CO-Alk \xrightarrow{e^-,H^+} Ar-COH-Alk$$

$$Ar-\underset{Ar-\underset{}{COH}-Alk}{COH}-Alk$$

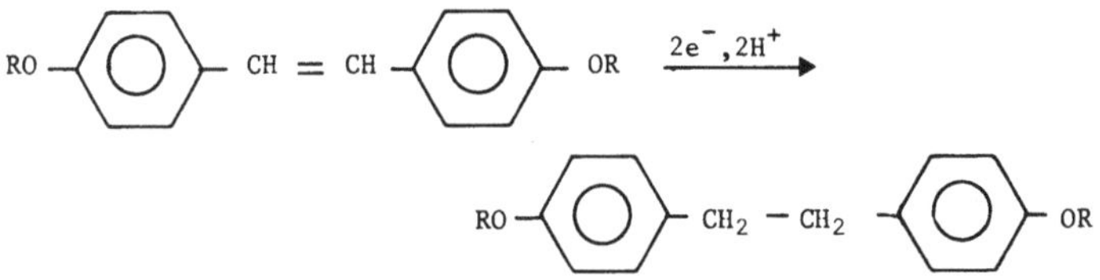

$$Ar-COH-Alk \xrightarrow{e^-,H^+} Ar-CHOH-Alk$$

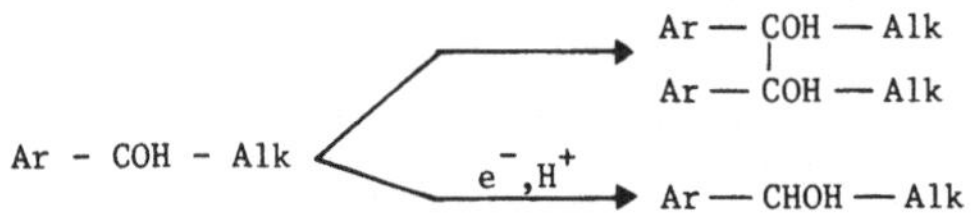

Bond Grouping	Example

C = N

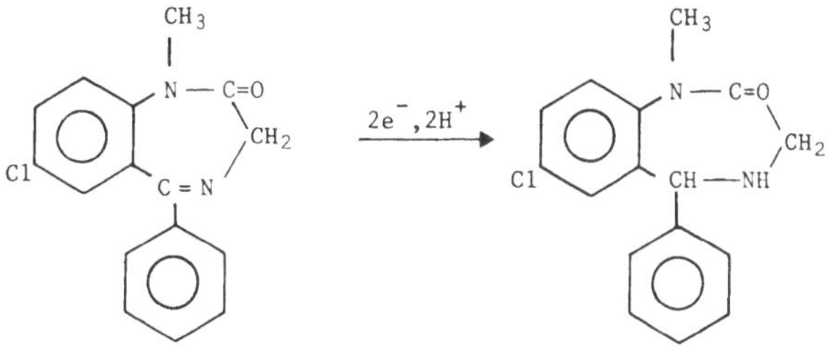

$$2e^-, 2H^+$$

N≡N

$$2e^-, 2H^+$$

$N \rightarrow O$

$$2e^-, 2H^+$$

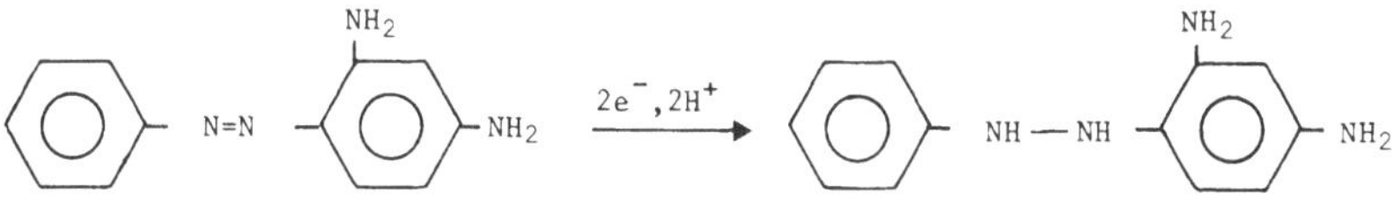

N = O

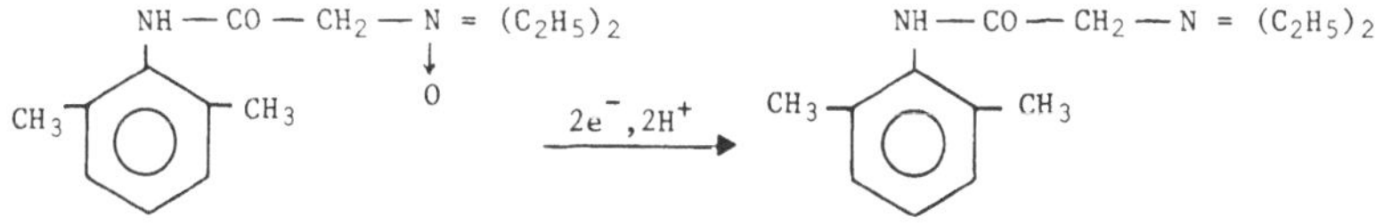

$$4e^-, 4H^+$$

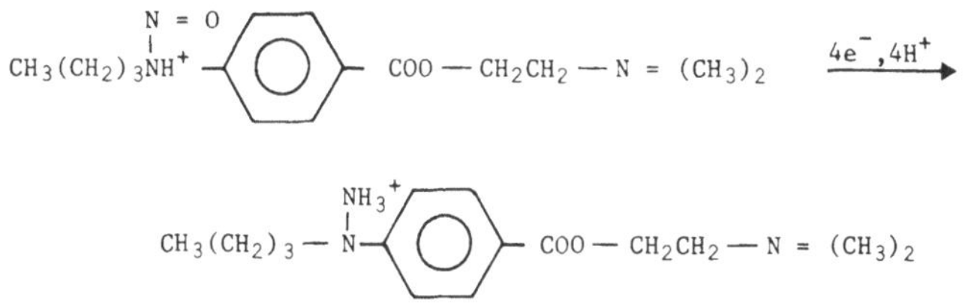

NO$_2$

$$4e^-, 4H^+$$

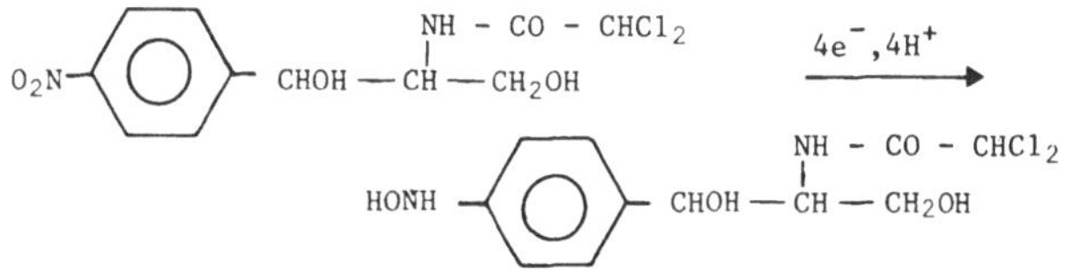

C = N
(heterocyclic)

$$2e^-, 2H^+$$

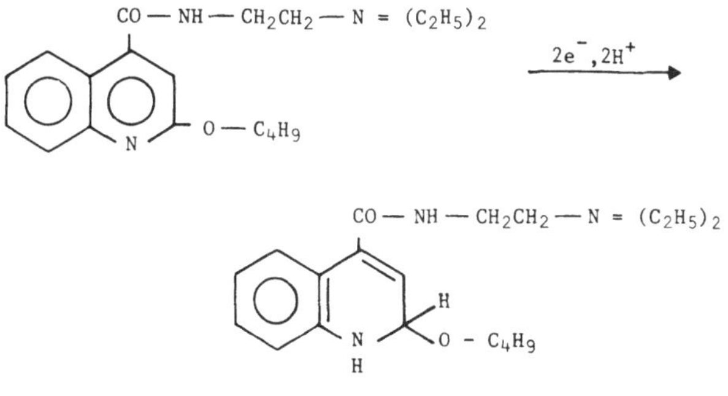

anodic oxidation but less data are known about the cathodic behavior
of organic substances in comparison to mercury electrodes. The
electrochemical activity of a substance is closely tied to the so-
called electrochemical domain (or voltage window)[3]: a substance is
electroactive in a given solution and at a given electrode if it can
be oxidized or reduced within the attainable voltage range.

     This voltage range is a function of the electrode material but
also of the solvent and of the supporting electrolyte: the potential
limit on the negative side is given by the reduction of the supporting electrolyte or of the solvent. Thus in a solution of 0.1 M
tetrabutylammonium perchlorate in anhydrous nitrobenzene the final
increase of the reduction current at -1.5 V vs. Ag/AgCl (at a rotated
Pt-electrode) is to be ascribed to the reduction of the quaternary
cation. On the positive side the increase of current probably corresponds to the oxidation of perchlorate. In general, the limit on the
positive side is given by the oxidation of the solvent, of the ions
of the supporting electrolyte or of the indicator electrode itself.
The latter is not only an anodic dissolution giving rise to ions but
also a dissolution of surface oxide films (e.g., with Pt). The
voltage limits for different electrode materials in various solvents
and supporting electrolytes are given in graphical form in Table 2.
The limiting currents of the substances to be determined (e.g., with
a rotated electrode) or peaks (in voltammetry with a stationary
electrode) are only obtained within the voltage window.

Table 2

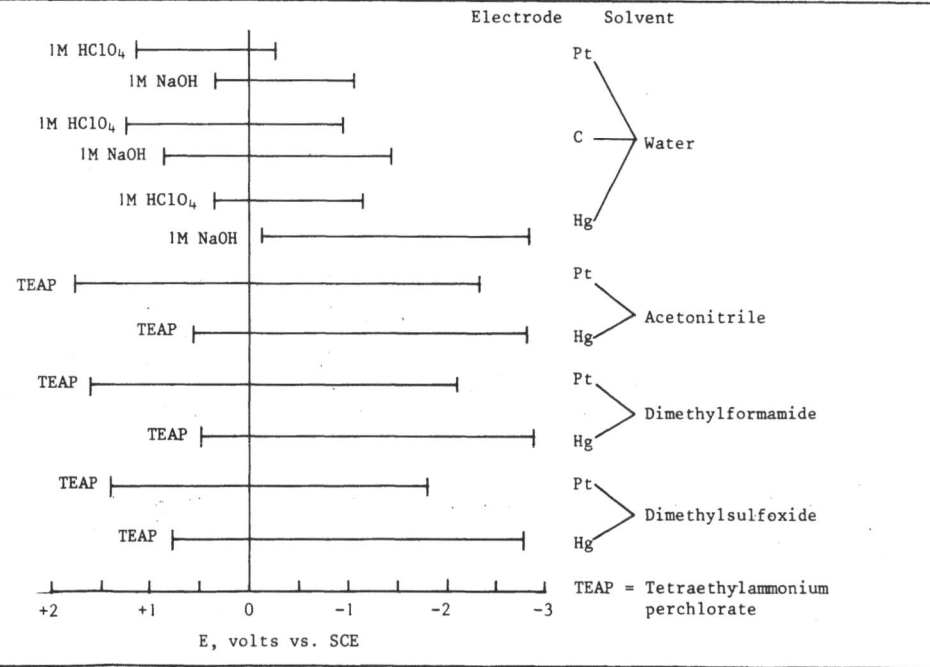

In contrast to the d.m.e. very strong residual currents must be taken into account. These are due not only to the charging of the electrode and further to the electrochemical reactions of traces of impurities present in the solution, but, as follows from cyclic voltammograms with a platinum electrode, different regions can be observed such as those of oxide formation and reduction and the regions of H_2 formation and its oxidation. For this reason a pretreatment[3] of a solid electrode is usually necessary; this can be performed either mechanically or by the so-called "cycling". As regards the cathodic limit, one has to bear in mind that in aprotic solvents the proton availability is much lower than in water and consequently the cathodic limit is extended.

The overvoltage of the electrode materials in aqueous solutions is demonstrated by the exchange current density i_o for the hydrogen evolution[3] reaction in 1 M H_2SO_4:

metal	$-\log i_o$ A/cm^2
Pd	3.0
Pt	3.1
Rh	3.6
Ni	5.2
Au	5.4
W	5.9
Ti	8.2
Cd	10.8
Tl	11.0
Pb	12.0
Hg	12.3

It follows from this that the exchange current for the reduction of hydrogen ions on platinum is about nine orders of magnitude larger than that on mercury ($7.3.10^{-4}$ A/$5.01.10^{-13}$A). Hydrogen overvoltage on mercury in aqueous solutions is much higher than on all abovementioned metals up to titanium. This is why in reductions on nonmercury electrodes such as Pt, Au etc., nonaqueous, aprotic solvents must be used. The situation with carbon electrodes (e.g., vitreous carbon) in the cathodic region is somewhat better so that reduction processes, e.g., in the aromatic ketones[4], can be investigated (cf. Figure 1).

Additionally, one thing has to be respected when comparing the electroactivity ranges of electrode-medium systems. It is the following difficult problem to be solved: Any comparison assumes the existence of a reference electrode which has the same potential in all media to be considered. This potential, however, is a function of the free energies of solvation of the two species which comprise the reference redox system; the free energies vary with the medium.

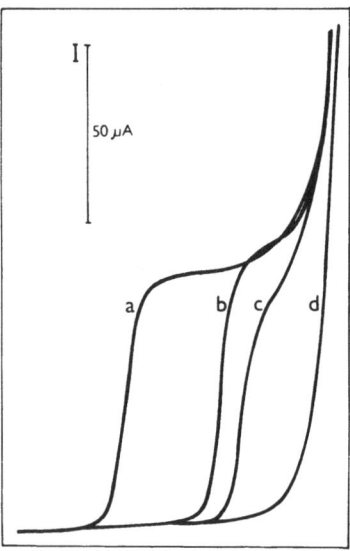

Figure 1. Current-voltage curves of 10^{-3}M benzil (a) benzoin (b) and
           acetophenone (c) in 0.1M LiCl with 50% acetone, rotated
           glassy carbon electrode.

A solution of the problem is a reference electrode whose constituents
are poorly soluble in any of the media. Such a couple is ferrocene-
ferrocinium whose potential does not depend on the solvent. Both
forms of this couple have a large volume; consequently the free
energies of solvation are very small. The application of such reference systems makes comparisons possible in which the potentials are
referred to a classical SCE in water.

     After all the above considerations and experimental data have
been taken into account one arrives at the following general conclusions regarding the electroactivity of organic substances at
non-mercury voltammetric indicator electrodes:

a)   the voltage windows are usually broader in nonaqueous solvents,
b)   the span of the voltage window is also strongly influenced by
     the current density,
c)   it follows from b that the data obtainable from literature on
     organic electropreparative reactions with non-mercury electrodes
     cannot be always used for assessing the possibilities of apply-
     ing these electrode materials in constructing electrochemical
     detectors or, in a more general way, for using them in voltam-
     metric measurements. Moreover, one has to consider here that
     electropreparative processes are often performed with controlled
     current,

d)   regardless of relatively few exceptions solid electrodes are
      more suitable for determining electroactive groups which are
      oxidized at the electrode surface,
e)   the reproducibility and, consequently, also the accuracy of
      current measurements with non-mercury electrodes is usually
      lower than that with mercury electrodes.

As regards available literature, the applicability of
Baizer's[5] "Organic Electrochemistry" is somewhat limited because of
its bias towards preparative electrochemistry. The same holds true
for the German textbook written by Beck[6]. More useful is in this
respect the monograph "Electrochemistry at Solid Electrodes" by
Adams[7]; however, this book was published 13 years ago. Interesting
data can be also found in Mann's monograph[8] on electrochemistry in
non-aqueous solvents.

In the following part of this paper I am going to present a
system of organic electroactive groups or of substances which can be
subjected to voltammetric analysis at solid electrodes forming thus a
base for construction of electrochemical detectors in HPLC. Since
electroactivity is the most important property of the substance to be
studied and we are interested here only in the analytical properties,
i.e. in the characteristical potentials (E_p or $E_{\frac{1}{2}}$) and in the shape
of the i_ℓ - E or i_p - E plot it will be not necessary to stress Adams
observation[7]: "Electrochemistry of organic oxidations is mainly
the chemistry of follow-up reactions" (of course, similar ideas hold
true with reductions). Nevertheless, electroactivity will be discussed here jointly with probable or proved interpretations of
mechanisms. This is done in contrast to Bond's view[9] who considers
interpretations of electrode processes and of their follow-up or
preceding reaction useless in analytical chemistry.

REDUCTIONS

1. Hydrocarbons[10,11]

Cathodic reductions can be carried out at platinum electrodes if
the measurement proceeds in aprotic media, e.g., in dimethylformamide
or acetonitrile and with tetraalkylammonium salts (perchlorates,
tetrafluoroborates, hexafluorophosphates) as supporting electrolytes.
A sufficient electron affinity of the hydrocarbon is necessary. The
process can be best exemplified by the reduction of 9,10-diphenyl-
anthracene in n-Bu$_4$NClO$_4$ at a Pt microelectrode:

$$(1)$$

The chemical follow-up reactions can be protonation, addition of electrophiles, homogeneous oxidation by another species present in the solution, dimerization and polymerization. Which of these processes will prevail depends on the structure of the reactant and on the composition of the solution.

In a step taking place at more negative potential a dianion results:

$$(2)$$

A certain redox symmetry exists in the oxidation/reduction behavior of hydrocarbons of this type: in benzonitrile as solvent a reversible oxidation of 9, 10-diphenylanthracene proceeds at the same electrode:

$$R \rightleftharpoons R^{\dotplus} + e \qquad \text{(from +0.04 V to +1.63 V)}$$
$$\text{cation radical}$$
$$R^{\dotplus} \rightleftharpoons R^{2+} + e \qquad \text{(from +0.30 V to +1.80 V)}$$
$$\text{dication}$$

$$(3)$$

In most cases this dication is much less stable than the radical anion. It rapidly reacts with a nucleophile present in the solution giving rise to a ∏-carbonium ion which is irreversibly reduced to the starting substance:

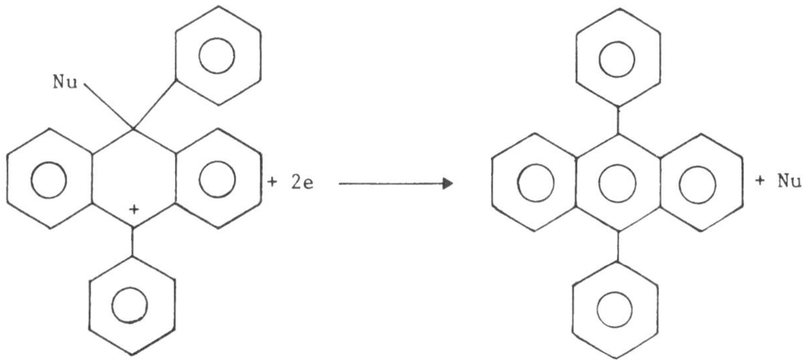

(4)

Nu⁻ = HO⁻, NH₂⁻, N⟨pyridine⟩ , CN⁻ etc.

2. Organic Halides[12]

Halogen-containing substances are usually electrolytically reduced on mercury electrodes but there is also a possibility of reducing them at lead, platinum or monel electrodes. The reduction of ethyl bromide at a lead cathode yields tetraethyl lead in a preparative reaction. Still, solid electrodes are usually not used for voltammetric measurements.

3. Nitro Compounds[13]

Nitrocompounds, in particular aromatic nitrocompounds are very easily reduced, especially in acid solutions: For this reason they most probably belong to those organic compounds for which solid electrode detectors can be constructed in combination with chromatographic separation techniques. This can be said although e.g. nitropyridines are separated by reversed phase chromatography and indicated by a mercury platinum electrode. The reduction mechanisms in aqueous solutions should be as follows:

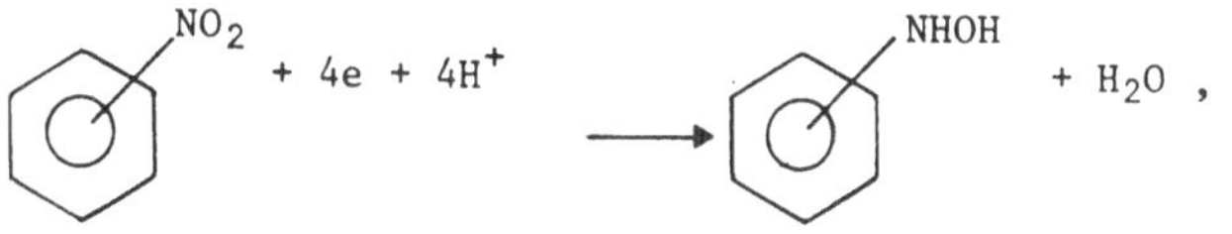

(5)

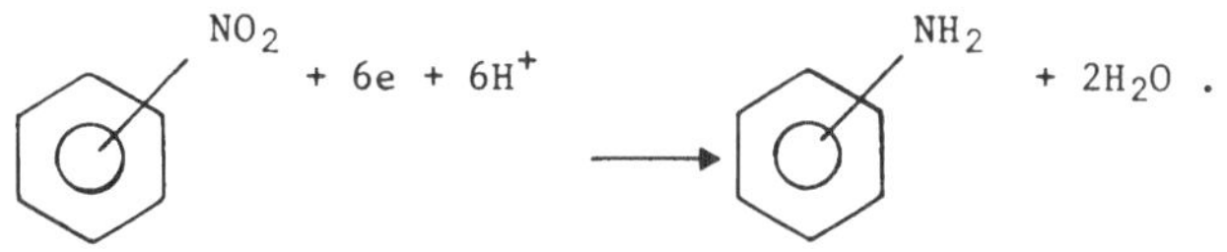

(6)

A similar technique has been recommended for determining p–nitroani–
line. It is known from electropreparative experiments that nitro
compounds can be reduced at tin cathodes.

4. Carbonyl Compounds[6]

Carbonyl compounds, especially aldehydes and ketones, regardless
of whether they are aliphatic, aromatic, heteroaromatic, or α,β–
unsaturated have been investigated mostly at mercury electrodes as
regards their reduction mechanism. It follows from literature,
however, that they can be reduced at copper, silver, platinum, tin,
nickel, aluminium, zinc or carbon cathodes. The mechanisms differ
strongly being a function of the electrode material, of the solution
composition and of the reducible carbonyl compound:

carbonyl compounds →
- glycols, dimers
- alcohols   primary, secondary
- hydrocarbons (in acid media)
- organometallics

The electron consumption varies from n = 1 in the case of dimer
formation to n = 2 if alcohols result, or n = 4 if the reduction
results in the formation of a hydrocarbon. In α,β–unsaturated
carbonyl containing compounds a 2–electron reduction takes place at
less negative potentials than the reduction of the carbonyl group.
In essence, a solid electrode can be applied for determining a car–
bonyl containing compound if the half–wave potential or the peak–
potential is substantially less negative than the final rise of
current corresponding to the reduction of the cations of the support–
ing electrolyte. The conditions at a carbon paste electrode are
rather favorable in this respect.

Even more convenient are the conditions for determining
quinones[14] at platinum or carbon solid electrodes because their
reversible 2–electron (or 2 one electron steps in aprotic media)
reduction proceeds at considerably positive potentials. This
property could be utilized in constructing an electrochemical detec–
tor for vitamin K derivatives:

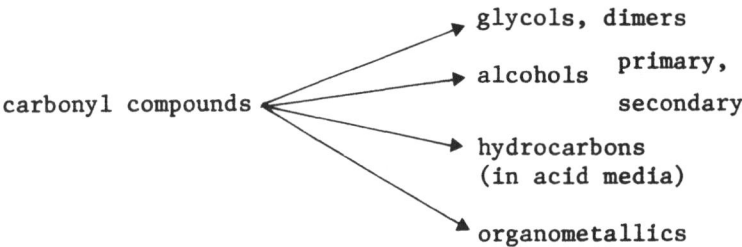

$$+ 2e + 2H^+ \quad (7)$$

$$R - phytyl \quad = \quad K_1,$$
$$difarnesyl \quad = \quad K_2,$$
$$H \qquad\quad = \quad K_3 \text{ (menadion)}.$$

ANODIC PROCESSES

With the help of solid electrodes numerous oxidation processes can be conveniently followed in voltammetry (or cyclic voltammetry) or organic compounds. This is why electrochemical detectors in chromatography often make use of platinum, carbon and other solid electrodes.

1. Hydrocarbons[15]

The key to the investigation of the anodic behavior of hydro-carbons lies in non-aqueous solvents since water-containing solutions lead to several parallel follow-up reactions and to a mixture of products in preparative experiments. In anhydrous acetonitrile the following direct oxidation mechanism is assumed at Pt electrode:

$$R\text{-}H \longrightarrow R\text{-}H^{+\cdot} + e \, , \tag{8}$$

$$R\text{-}H^{+\cdot} \longrightarrow R\cdot \Big\langle {}^{Nu}_{H} \quad , \tag{9}$$

i.e. a cation radical and a nucleophile,

$$R\text{-}H^{+\cdot} + B \longrightarrow R\cdot + BH^{+} \, , \tag{10}$$

or a cation radical and a base

The oxidation potential of $R\cdot$ or $R\cdot{\textstyle<}^{Nu}_{H}$ is less positive than that of R-H; hence in the second oxidation step

$$R\cdot\Big\langle {}^{Nu}_{H} \longrightarrow R\text{+}\Big\langle {}^{Nu}_{H} + e \, , \qquad ECE \tag{11}$$

or

$$R\cdot \longrightarrow R^{+} + e. \tag{12}$$

The cations are converted to neutral particles by a second chemical reaction with a nucleophile or with a base. (It cannot be excluded that

$$R-H^{+}_{\cdot} + R-H^{+}_{\cdot} \rightleftharpoons R-H + R-H^{2+} \qquad EC.)$$
$$\text{dication}$$

An indirect oxidation mechanism can be also assumed via the anion of the supporting electrolyte:

$$X^{-} \longrightarrow X^{\cdot} + e$$
$$R-H + X^{\cdot} \longrightarrow R^{\cdot} + H-X \qquad (13)$$

(e.g., a C-H bond in the benzylic position of an alkyl-aromatic hydrocarbon).

1.1. <u>Aromatic Hydrocarbons</u>[10] A well-developed 2e wave results at a rotated Pt anode (0.5 M NaClO$_4$, acetonitrile)

Compound	$E_{1/2}$ (vs. Ag/Ag^{+})
Benzene	2.00
Toluene	1.98
o-Xylene	1.57
m-Xylene	1.58
p-Xylene	1.56
Hexamethylbenzene	1.16
Naphthalene	1.34
Anthracene	0.84
Phenanthrene	1.23
Biphenyl	1.48

An erroneous 2e mechanism has been presented by Lund. For less easily oxidizable hydrocarbons acetonitrile with tetrabutylammonium tetrafluoroborate as supporting electrolyte is used.

1.2 <u>Aliphatic and Alicyclic Hydrocarbons</u>[10] The anodic limit in perchlorate is not sufficient. The oxidation only proceeds in tetrafluoroborate or in hexafluorophosphate. There is a good correlation between $E_{1/2}$ at a rotated electrode and the ionization potentials. The diffusion currents correspond to the uptake of 2 electrons.

It holds for simple aliphatic hydrocarbons:

$$R-H \xrightarrow[-H^{+}]{-2e} R^{+} \xrightarrow[H_2O]{CH_3CN} RNHCOCH_3$$
$$\text{N-substituted acetamide} \qquad (14)$$

Compound	$E_{1/2}$ V
n–Alkanes	3.4
Ethylene	2.90
1–Alkenes	2.7 - 2.8
2–Alkenes	2.2 - 2.3
1,4–Cyclohexadiene	1.3
1,3,5–Cycloheptatriene	1.13
Cyclooctatetraene	1.2

The follow-up reactions in all hydrocarbon oxidations can be sub-stitutions, additions and coupling with eliminations or additions.

2. Carboxylic Acids[6]

The anodic oxidation of a carboxylic acid in the form of its alkali metal salt is the oldest and still very useful method of preparative organic electrochemistry (Kolbe reaction):

$$2RCOO^- \longrightarrow R-R + 2CO_2 + 2e \qquad (15)$$

However, there are not enough data available in the literature about the voltammetric curves resulting at solid electrodes and correspond-ing to the oxidation of carboxylic acids.

3. Amines

3.1. Aliphatic and Benzyl Amines[16] Primary amines are easily oxidized at platinum electrodes, the first step being as follows:

$$R-CH_2NH_2 \longrightarrow R-CH_2\overset{+\cdot}{N}H_2 + e \overset{-H^+}{\rightleftharpoons} \dot{R}CH_1NH_2 \qquad (16)$$

3.2 Aromatic Amines Primary Amines: Anilines in aqueous media: p-substituted anilines react in the following way: a coupling reaction involving elimination of the para substituent to give

4'-substituted 4-aminodiphenylamines:

(17)

With   p-chloraniline, :   1e pathway;
                        p-anisidine,
                        p-phenetidine,

with   aniline,   2e pathway.
                        o-toluidine

p-Chloraniline:

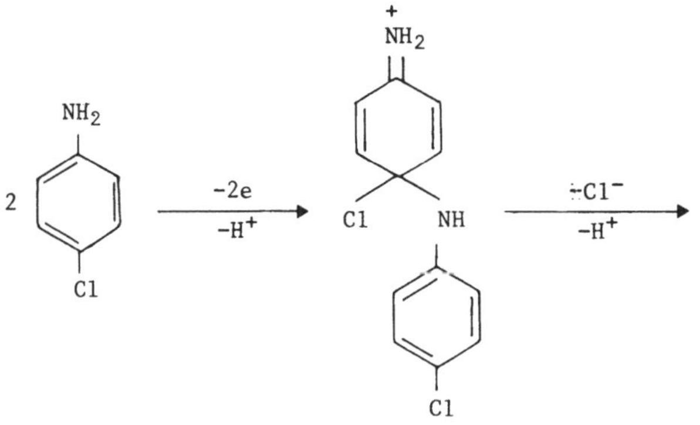

(18)

a 1e process

for aniline

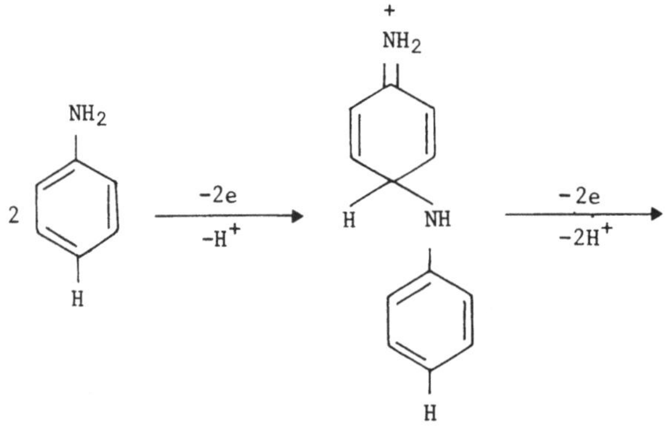

(19)

a 2e process

A competing reaction, the benzidine formation, occurs with substrates non-substituted in the para position.

2,4,6-tri-tert-butylaniline (from cyclic voltammetry):

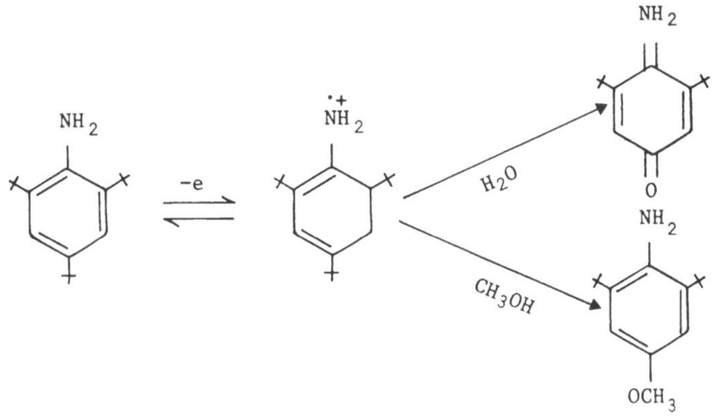

(20)

Secondary Aromatic Amines

N-Methylaniline reacts in aqueous solutions in the following way:

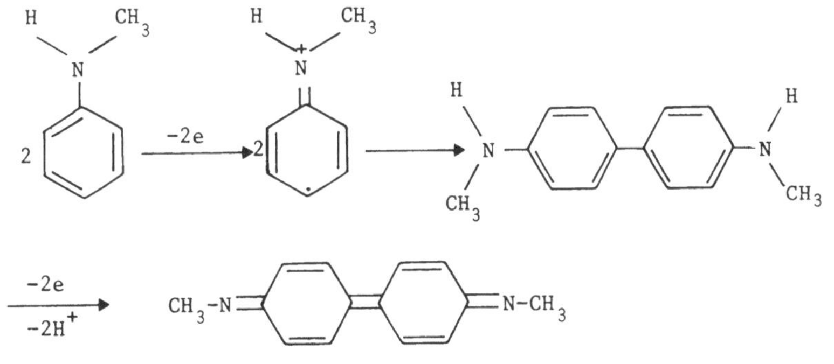

(21)

oxidized dimethylbenzidine

Diphenylamine also loses 2 electrons:

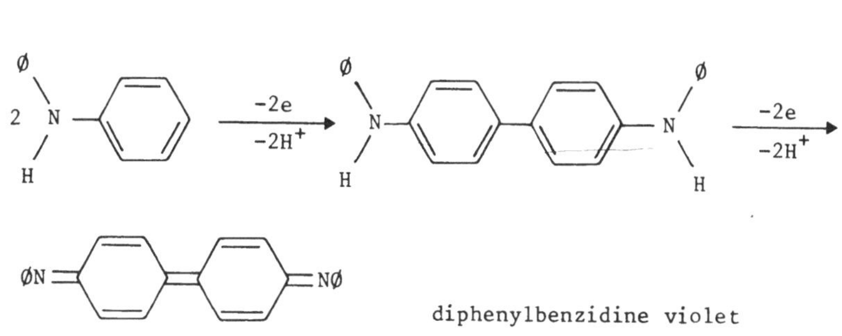

(22)

diphenylbenzidine violet

Tertiary Aromatic Amines

N,N-dimethylaniline (aqueous solution) first undergoes a single
electron reversible reaction:

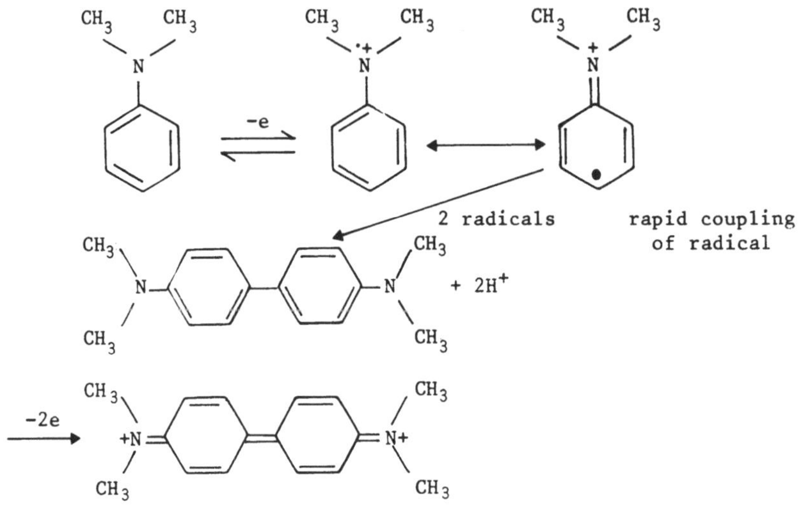

(23)

Phenylenediamines

Here, only a simplified version for aqueous solutions is shown:

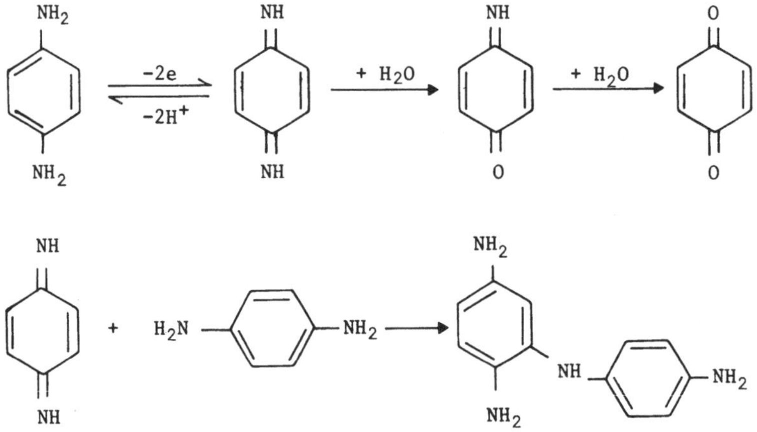

(24)

(25)

Aminophenols

A typical EC process in aqueous solutions has been observed which is
used as a test system for evaluating the applicability of voltammetric techniques:

$$(26)$$

k_h measured as a function of pH by different methods

4. Oxygen-containing Compounds[17]

This group involves phenols, hydroquinones and catechols, aromatic ethers and esters and, finally, alcohols.

Phenols and their Derivatives Simple phenols: detailed studies exist on the anodic oxidation of 2,6-di-tert-butyl-p-cresol in acetonitrile. The interpretation results in the following oxidation mechanism:

$$(27)$$

Somewhat different is the course of electrooxidation of 2,4,6-triphenylphenol (in $CH_3COOH - CH_3COONa$):

$$(28)$$

Hydroquinones and Catechols The anodic oxidation of hydroquinones in buffered aqueous solutions is well known. It proceeds reversibly via a radical species the charge of which depends on the proton availability in the solution. The product of the overall 2-electron oxidation is the corresponding quinone.

In aprotic solvents such as acetonitrile, constant potential electrolysis yielded benzoquinone as final product whereas in cyclic voltammetry on a platinum anode a one-electron step can be observed the product here being quinhydrone.

This class of compounds includes several important biologically active substances, i.e., catecholamines, adrenaline, noradrenaline, dopamine, etc. The electrochemical behavior found here can be exemplified with adrenaline in acid solutions:

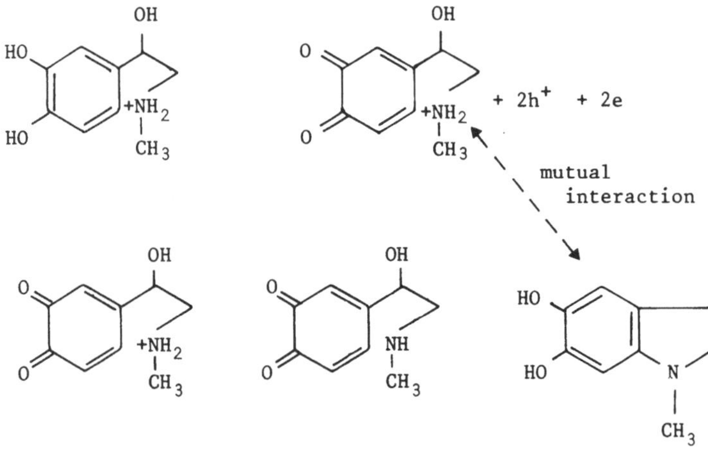

(29a)

mutual interaction

(29b)

Aromatic Ethers and Esters These substances (e.g., 4-methoxyphenol or 4,4'-oxydiphenol) are oxidized in water-containing solutions according to the following scheme:

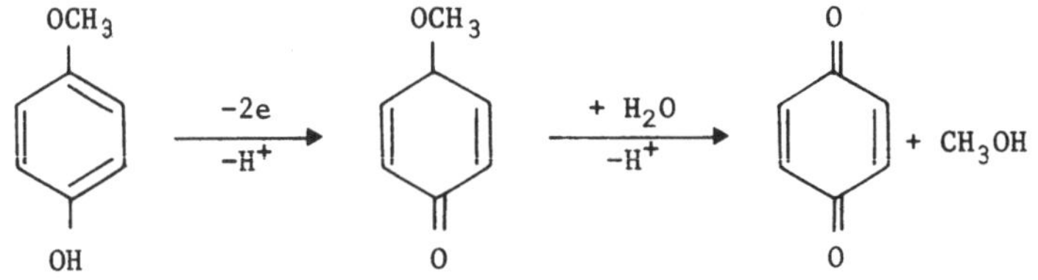

(30)

This reaction proceeds quantitatively. A tocopherol model can also be looked upon as an ether. The following EC mechanism has been observed:

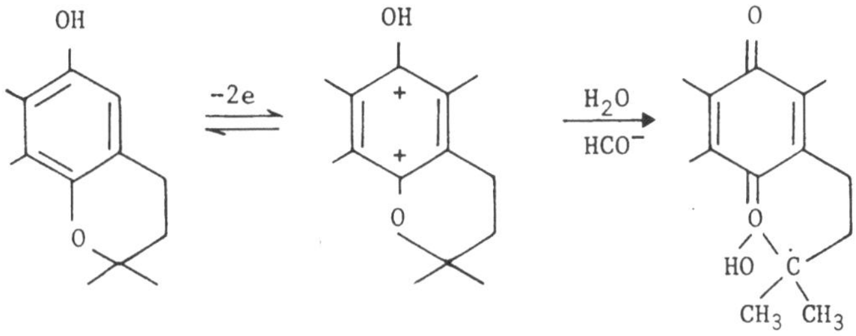

(31)

The above dication is stable in acetonitrile. The primary product in dry acetonitrile is somewhat different. It bears only one positive charge:

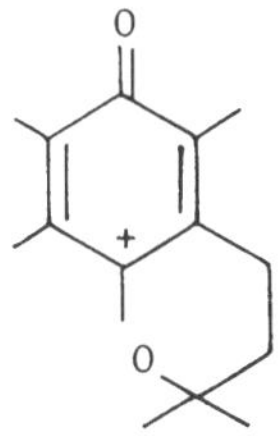

Methyl and acetyl transfers from oxygen have been observed, e.g., with 9-anthranol derivatives:

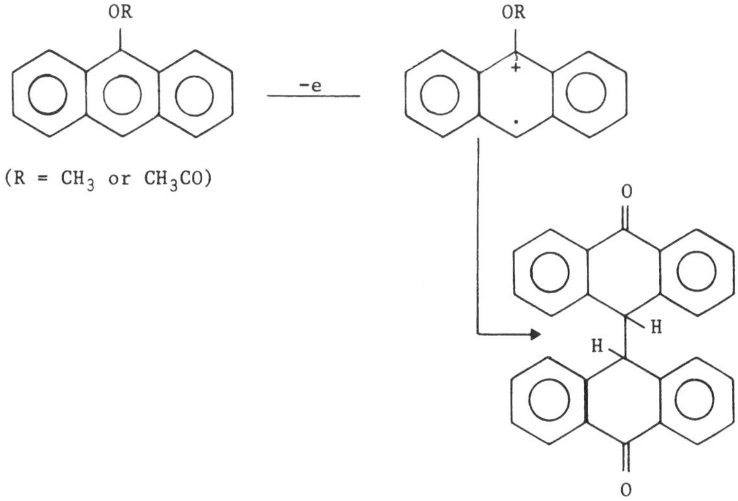

(R = CH_3 or CH_3CO)

(32)

__Alcohols__ Carbinols give waves and peaks at Pt, PbO_2^- and C electrodes in acetonitrile as may be shown with substituted benzylalcohols:

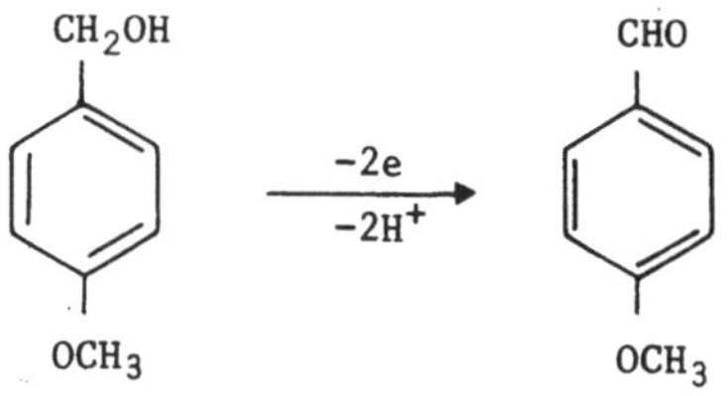

$$
\text{(33)}
$$

5. Sulfur-Containing Compounds[18]

Here, one must distinguish between electrooxidations in absence and in presence of water. At a platinum anode the oxidation leads to the formation of sulfonium ions:

$$
CH_3-S-CH_3 \xrightarrow{-e} CH_3-\overset{+\cdot}{S}-CH_3 \longrightarrow
$$

$$
\xrightarrow[-e]{-H^+} CH_3-\overset{+}{S}=CH_2 \longrightarrow CH_3-\overset{+}{S}-CH_2 \qquad \text{(34)}
$$

$$
CH_3-\overset{+}{S}-CH_2 + (CH_3)_2S \rightleftharpoons CH_3-S-CH_2-\overset{+}{S}\begin{smallmatrix}CH_3\\ \\CH_3\end{smallmatrix} \qquad \text{(35)}
$$

In presence of even 1% water a different mechanism is operative:

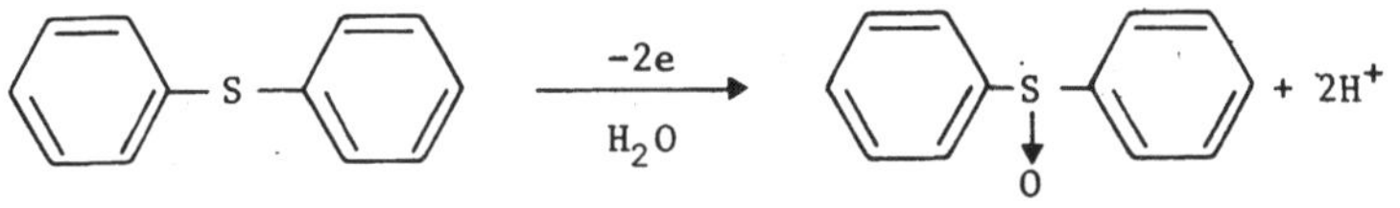

$$
\xrightarrow[H_2O]{-2e} \qquad + 2H^+ \qquad \text{(36)}
$$

diphenyl sulfoxide

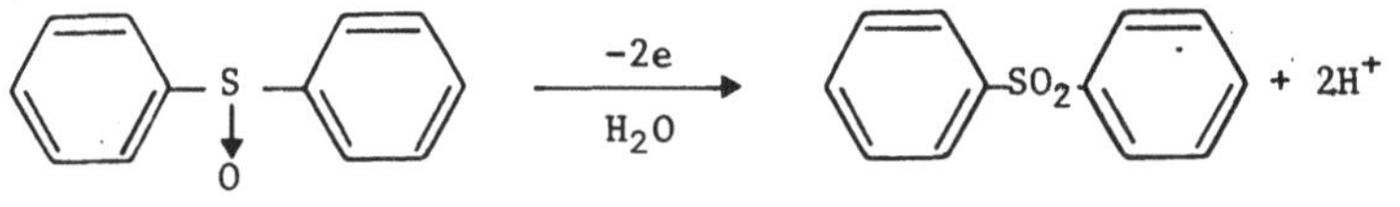

$$
\xrightarrow[H_2O]{-2e} \qquad + 2H^+ \qquad \text{(37)}
$$

diphenyl sulfone

6. Nitrogen- and Sulfur-Containing Heterocyclics

Only a few examples of heterocyclics will be presented since this group includes a large number of compounds. Their reduction often proceeds at relatively negative potentials unsuitable for the application of a solid electrode with a low hydrogen overvoltage.

Therefore the oxidation processes of these compounds at platinum or carbon electrodes will be presented which are, or could be, used for constructing an electrochemical detector.

3,5-Difunctional 1,4-dihydropyridine derivatives[19] are relatively easily oxidized at platinum electrodes (e.g., rotated platinum). In solutions containing a mixture of acetonitrile and water (1:1) and 0.05 M LiClO$_4$ a 1 electron anodic wave results with 2,4,4,6-tetramethyl-3,5-dicyano-1,4-dihydropyridine (I) and a 2 electron wave with 2,6-dimethyl-3,5-dicyano-1,4-dihydropyridine (II):

The overall oxidation mechanism is as follows:

(38)

The anodic half-wave potential lies in the vicinity of +1 V. The existence and the life-time of the primary radical cations could be proved by EPR spectroscopy or by cyclic voltammetry. Its stability is increased by methyl groups in position 4 and by substitution in position 1.

The electrochemical behavior of the reduced form of NADH, i.e. nicotinamide adenine dinucleotide[20] bears a strong resemblance to the above behavior. Further the oxidation of purine and of its derivatives at pyrolytic graphite electrodes must be mentioned:

Purine   Adenine   Cytosine

Pyrimidine   Uric acid

Dryhurst worked out a voltammetric method for the simultaneous determination of guanine and guanosine in mixtures. A similar method based both on cathodic and anodic waves (a graphite electrode) has been published for allopurinol and uric acid.

Allopurinol   Uric acid

An anodic oxidation has been also found with theobromine and coffeine.

In case of guanine, xanthine, hypoxanthine and adenine a separation on a chromatographic column has been suggested followed by anodic determination on a silicone-rubber based carbon electrode.

A great deal of attention has been directed toward the voltammetric investigation of tricyclic aromatic heterocycles with two heteroatoms such as phenothiazine[21], phenazine[22] and thianthrene[23] at platinum electrode (in anhydrous acetonitrile)

$$\text{(39)}$$

$$\text{(40)}$$

The substituted phenothiazines play a considerable role in biomedical analysis.

7. Pharmaceuticals and Biologically Active Substances

There is a special group of organic compounds which is rather non-homogeneous as regards their chemical structure, and, consequently, their chemical and electrochemical behavior strongly differ. They are all the biologically active substances which are in most cases used as pharmaceuticals but they may also play a role in toxicology or, perhaps in environmental chemistry. Few examples will also be mentioned here. The classification and grouping of compounds is performed here mostly according to physiological properties:

Alkaloids The 2-electron anodic oxidation of morphine[24] at a carbon paste (or platinum) electrode in 0.2 M NaOH is described by the following system of equations as postulated by the authors:

$$\text{(41)}$$

The anodic activity has been only observed in those morphine derivatives exhibiting phenolic functions such as psuedomorphine, dihydro-
morphine, dihydromorphinone, nalorphine, apomorphine etc. On the
other hand codeine, ethylmorphine, thebaine, benzylmorphine or dia-
cetylmorphine (heroine) are inactive.

Analgesics[25] Some analgesics exhibiting a phenolic character or
being amines are oxidized at glassy carbon or silicone rubber based
graphite electrodes as well as at a rotated platinum disk. The
following compounds are to be mentioned:

paracetamol CH_3CONH —⟨ ⟩— OH

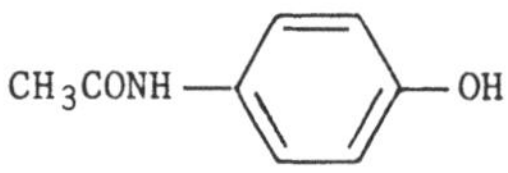

phenacetine CH_3CONH —⟨ ⟩— OC_2H_5

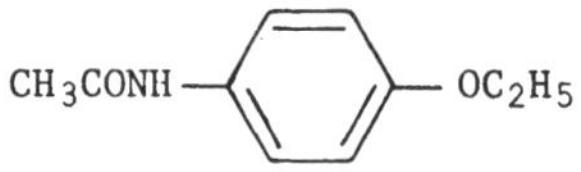

p-aminosalicylic acid $HOOC$—⟨ ⟩— OH

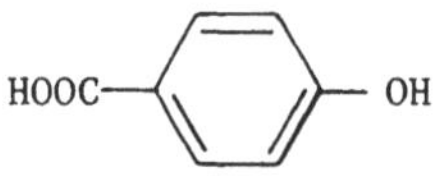

amidopyrine

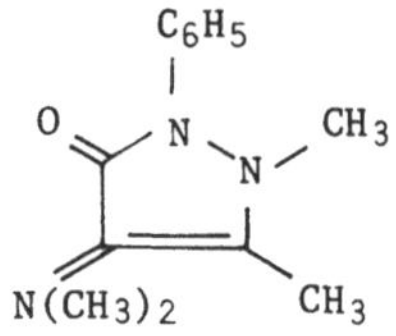

$$
\begin{array}{c}
C_6H_5 \\
| \\
O = \underset{}{N} \diagdown \underset{N}{} \diagup CH_3 \\
\diagup \diagdown \\
N(CH_3)_2 \quad CH_3
\end{array}
$$

Psychotropic Substances[26,27] The modern psychotropic substances
include a group of compounds derived from dibenzocycloheptane such as
amitriptyline (5-(3-dimethylaminopropylidene)-10,11-dihydro-5H-
dibenzo[a,d]cycloheptene) or of its heterocyclic analogues. In
acetonitrile (0.1 M LiClO$_4$) they give an anodic wave at a Pt electrode; the wave can be ascribed to the following reaction

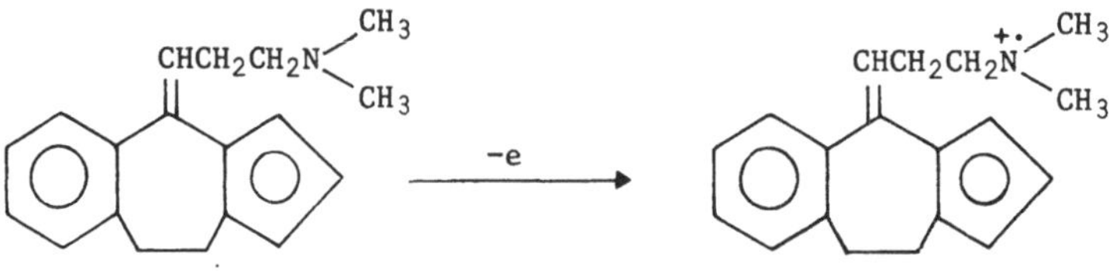

$$\text{(42)}$$

This electrode reaction is followed by the loss of a proton and of a further electron (ECE), i.e. the wave corresponds to a 2e oxidation.   In the sulfur-containing analogue (the so-called prothiaden) adsorption phenomena make the behavior more complicated.

Hormones[28]   The phenolic substances which can be oxidized at a glassy carbon electrode include some hormones.   After a HPLC separation the following substances have been determined: oestriol, 17β-oestradiol, oestrone, diethylstilboestrol, dienoestrol and hexoestrol.   The non-dissociated forms are oxidized in a 2-electron process, the phenolate in a 1-electron step only.   The processes are as follows:

oestriol, 17β-oestradiol, oestrone:

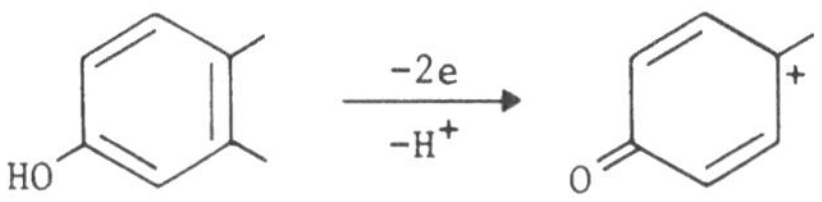

diethylstilboestrol, dienoestrol:

E_p = +0.54 V:

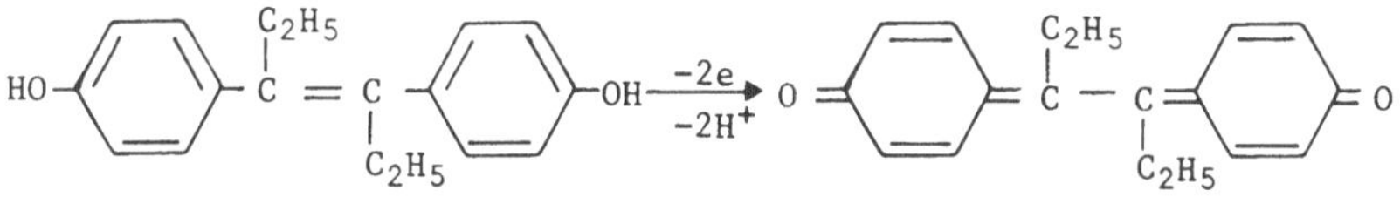

E_p = +0.90 V:

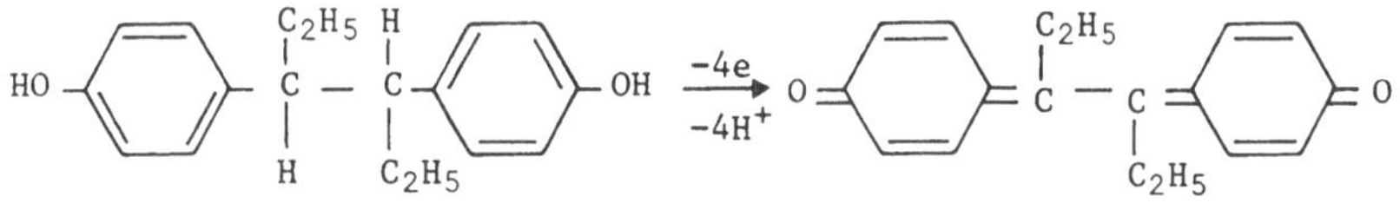

In the mycotoxine zearalenone a cis and a trans form exist. Both give an anodic wave at glassy carbon anodes (acetonitrile : water 45/55 in 0.05 M $LiClO_4$).   The trans form is oxidized with E_p = +1.05 V whereas cis with E_p = +1.02V.

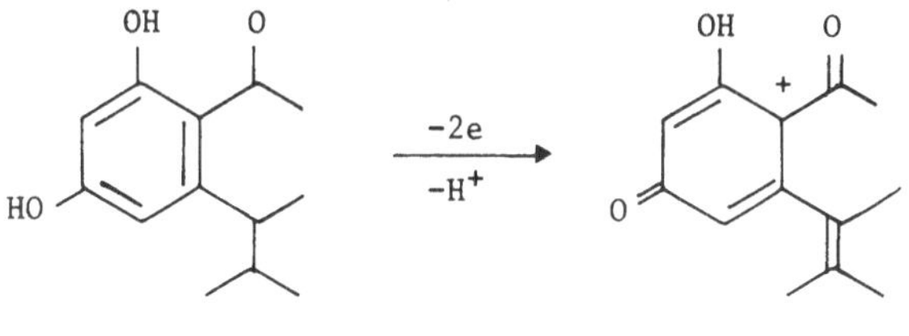

(45)

The cis form is less harmful.

Conclusions

Based on the material presented in the preceding pages one can
assume that in spite of the unsuitability of non-mercury electrodes
in most cathodic reductions (as compared to mercury) a considerable
number of organic substances can be oxidized at these electrodes and
the resulting voltammetric curves can be used in quantitative analysis. The compounds of the following classes can be thus particularly
conveniently determined by anodic processes:

    aromatic hydrocarbons,
    aliphatic and alicyclic hydrocarbons, alkenes,
    carboxylic acids,
    aliphatic and benzyl amines,
    aromatic amines,
    aminophenols,
    phenols, hydroquinones and catechols,
    aromatic ethers and esters,
    alcohols,
    sulfur-containing compounds (sulfides, disulfides)
    nitrogen- and sulfur-containing heterocycles.

REFERENCES

1. CRC Handbook Series in Organic Electrochemistry, Volumes I-IV
       (L. Meites, P. Zuman, eds.), CRC Press, Inc., Cleveland, Ohio
       and West Palm Beach, Fla. (1978).
2. H. Hoffmann and J. Volke, in "Electroanalytical Chemistry", Vol.
       10 in "Advances in Analytical Chemistry" (H.W. Nürnberg,
       ed.), J. Wiley and Sons, London p. 287 (1974).
3. D. T. Sawyer and J. L. Roberts, Jr., "Experimental Electro-
       chemistry for Chemists", J. Wiley and Sons, New York p.
       (1974).
4. J. Weber and J. Volke, Electrochim.Acta, 24:113 (1979).
5. "Organic Electrochemistry" (M.M. Baizer, ed), M. Dekker, Inc.,
       New York p.1072 (1973).
6. F. Beck, "Elektroorganische Chemie, Grundlagen und Anwendungen",
       Akademie Verlag, Berlin (1974).
7. R. N. Adams, "Electrochemistry at Solid Electrodes", M. Dekker,
       Inc., p. 402, New York (1969).
8. C. K. Mann and K. K. Barnes, "Electrochemical Reactions in
       Nonaqueous Systems", M. Dekker, Inc., p. 560, New York
       (1970).
9. A. M. Bond, "Modern Polarographic Methods in Analytical
       Chemistry", M. Dekker, Inc., p. 518, New York (1980).
10. L. Eberson in 5, Chapter XII, p. 447; this is a general review.
11. J. Phelps, K. S. V. Santhanam and A. J. Bard, J.Am.Chem.Soc.,
       89:1752 (1967). Most recent results are to be found the the
       series of very profound papers by V.D. Parker and coworkers,
       appearing in Scand.Chim.Acta.

12. M. R. Rifi in 5, p.301; most data, however, refer to mercury cathodes, only exceptionally lead, platinum and Monel have been used.

13. Reviewed by H. Lund in 5, p. 316; more recent data to be found in 6.

14. L. Meites and P. Zuman, Electrochemical Data, Part I, Vol. A., J. Wiley and Sons, p. 174, New York (1974).

15. H. Lund, Acta Chem.Scand., 11:1323 (1957); M. E. Peover in "Electroanalytical Chemistry", (A.J. Bard, ed.), Vol. 2, M. Dekker, New York, p.1, (1967) (Review).

16. Anodic processes of amines are described in most detail either in 7 or in a special chapter in 5. This chapter (which is more recent) has been written by V. D. Parker who published the most fundamental papers in this field.

17. The same holds true as in reference 16.

18. V. D. Parker in reference 5, p. 551.

19. V. Skála, J. Volke, V. Ohánka and J. Kuthan, Coll.Czech.Chem. Comm., 30:2632 (1975).
    J. Klíma, A. Kurfürst, J. Kuthan and J. Volke, Tetrahedron Letters No. 31:2725 (1977).
    J. Ludvík, J. Klíma, J. Volke, A. Kürfurst and J. Kuthan, J.Electroanal.Chem., in press, (1982).

20. W. T. Bresnahan, J. Moireaux, Z. Samec and P. J. Elving, Biochem.Bioenergetics, 7:125 (1980). J.Electroanal.Chem., 116:125 (1980).

20b. L. Underwood and R. W. Burnett, in "Electroanalytical Chemistry", (A.J. Bard, ed.), Vol. 6, M. Dekker, p.1, New York (1973), a good review.

21. J.-P. Billon, Ann.Chim., 7,196 (1962).

22. R. F. Nelson, D. W. Leed, E. T. Seo and R. N. Adams, Z.Anal.Chem., 224:184 (1967).

23. V. D. Parker and L. Eberson, J.Amer.Chem.Soc., 92:7488 (1970).

24. B. Proksa and L. Molnár, Anal.Chim.Acta, 97:149 (1978).

25. H. K. Chan and A. G. Fogg, Anal.Chim.Acta, 105:423 (1979).
    E. Pungor, Z. Fehér and G. Nagy, Acta Chim. Acad.Sci.Hung., 70:207 (1971).
    C. M. Shearer, K. Christenson and G. J. Papariello, J.Pharm. Sci., 61,1627 (1972).

26. J. Volke, M. M. Ellaithy and V. Volková, J.Electroanal.Chem., 60:239 (1975).

27. R. Rucki et al., J.Pharm.Sci., 67:1297 (1978).

28. M. R. Smyth, C. G. B. Frischkorn, Z.Anal.Chem., 301:220 (1980).

TENSAMMETRY IN COMBINATION WITH ADSORPTIVE ACCUMULATION

OF SURFACE ACTIVE COMPOUNDS ON THE ELECTRODE SURFACE

R. Kalvoda

The J. Heyrovský Institute of Physical Chemistry
and Electrochemistry
Czechoslovak Academy of Sciences, Prague 1

Recent substantial improvements in polarographic stripping analysis have resulted from the automation of the whole procedure where all necessary operations are programmed and the strictly reproducible conditions led to higher precision and accuracy. This gain has been assisted by the development of new types of sensors - mainly the static mercury drop electrode (SMDE). This electrode serves either as a hanging Hg electrode or as a dropping one. It can be used also as a dropping electrode "with constant surface area" after the drop is formed. The polarization of the dropping electrode in the moment when the surface area is just constant contributes to the decrease of charging current and the achievement higher sensitivity. At this static mercury drop electrode a needle valve is opening and closing the mercury flow through the capillary. Thus the drop is stopped at a determined instant of its growth. The movement of the needle is controlled from the polarograph. Using the SMDE as a dropping electrode a tapper dislodges the old mercury drop periodically from the capillary. The end of this glass capillary has a spindle shaped inner space [1]. The spindle shaped capillary yields drop-times of high reproducibility. In history there were many attempts to develop such a type of electrode. One of them was described by Gokhshtein et al. [2], and produced by the Academy of Sciences of the USSR. New types, where the drop formation and its size are controlled electronically are now on the market (PARC USA [3], Tesla Laboratorní prístroje, CSSR).

A static mercury drop electrode (in combination with the automated polarograph PA3 produced by Tesla Laboratorní prístroje Prague) was used by the present author for the further investigation of adsorptive phenomena. This work was started many years ago in connection with the study of the behavior of surface-active compounds by

means of oscillographic polarography with alternating current.[4]
Studying the oscillographic behavior of benzophenone and related
compounds[5] it was observed that the incision on the dE/dt=f(E)
curve becomes deeper if the electrode is first held for some brief
time at a constant potential where adsorption of the compound takes
place (Figure 1). Similar effects were also observed with some metal
ions in solutions containing complex forming compounds or ions adsorbable at the electrode surface. The adsorptive accumulation effect
was later used by many authors for the determination of various
compounds; with different polarographic techniques being used as the
stripping step. A survey is given in reference 6.

The aim of the present work is the study of the adsorptive
accumulation of surface active compounds on the electrode surface in
polarographic stripping analysis, where, during the stripping process, the desorption peak is recorded using differential pulse polarography (DPP).

RESULTS AND DISCUSSION

For adsorptive accumulation in stripping polarography there are
suitable compounds yielding well-developed tensammetric DPP peaks in
the concentration range from 10^{-5}M to 10^{-6}M at the dropping Hg electrode. Such strongly adsorbable compounds usually have an adsorption
coefficient of about $10^{-3}M^{-1}$ or higher in the respective supporting
electrolyte. The adsorption can be influenced by proper choice of
supporting electrolyte or variations in its concentration as in the

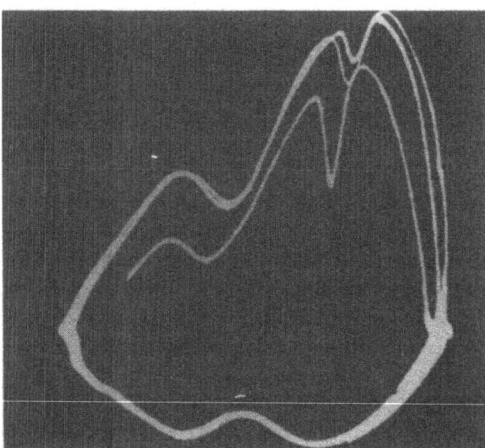

Fig. 1. dE/dt = f(E) curve of benzophenone at polarization with
single cycles of a.c. current. 4.10^{-5}M benzophenone in 1M
H_2SO_4 (at the first cycle the curve with the biggest
indentation was obtained).

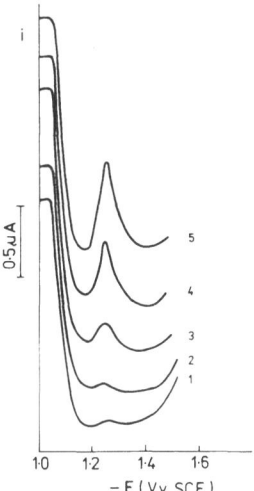

Fig. 2.   Stripping analysis with adsorptive preconcentration: 8 x
          10^{-8}M codeine in 5M NaOH. Curve 1, normal DPP record
          without preconcentration; curve 2, during preconcentration
          step unstirred, accumulation time 60 s; curve 3, as curve 2,
          but accumulation time 180 s; curve 4 and 5, during
          preconcentration step stirred, accumulation time 20 s and
          40 s respectively. Accumulation potential –1.0 V (SCE), scan
          rate 20 mV/s.

case of 5M NaOH (Figure 2) where sometimes the salting out effect of
the compound contributes to an increase of the capacitive phenomena.

      To find out the optimal conditions for the adsorptive accumu-
lation of the compound under investigation, is recommended as a first
step to record the DPP tensammetric curve over the whole accessible
potential range using the SMDE as a dropping electrode. The optimal
supporting electrolyte, the optimal adsorption potential for the
accumulation as well as all other usual parameters can be determined
in these experiments. The accumulation effect of the compound under
investigation is then tested using the hanging mercury drop electrode. After recording the DPP curve with the tensammetric peak, the
same curve is rerun but after adsorptive accumulation at the most
suitable potential. An increase in the peak height is the proof of
accumulation (Figure 3). The procedure in tensammetric stripping
experiments is as follows: After deaeration of the solution, the HMD
is formed (using SMDE) and the constant voltage applied for a given
time interval. The accumulation proceeds in unstirred as well as in
stirred solutions; in the second case, of course, a shorter accumulation time is sufficient (Figure 2). After the accumulation time
had elapsed, the record of the curve is automatically started (after
a quiescent period lasting several seconds). The potential scan

takes place from the accumulation potential to more negative values
while the desorption peak is recorded. For every measurement a new
drop of the SMDE was formed. The peak height in this type of stripping analysis depends on the duration of the preconcentration period
and reaches a limiting value. A typical concentration dependence of
the peak height is in Figure 4. Thus in the analysis of practical
samples the best method is the use of a calibration curve. It is
necessary to perform the stripping procedure also with the blank
supporting electrolyte, because, mainly at a longer accumulation
period, impurities from the solution are adsorbed at the electrode
surface and can affect the adsorption of the studied compound (e.g.,
through competitive coverage of the electrode surface or through
other interference).

    The results obtained are summarized in Table 1. The given
concentration does not denote the detection limit but rather a concentration at which a well developed peak was obtained under reasonable signal/noise conditions. Measurements in less concentrated
solutions can be performed using longer accumulation time, stirring
the solution and using higher amplification in the polarograph providing a good signal to noise ratio is maintained. In this case
strict limitations on the purity of the base solution must be pre-
served.

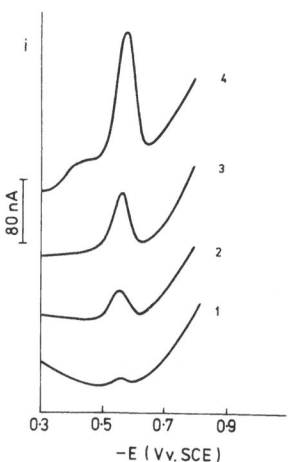

Fig. 3. The influence of precedent adsorptive accumulation on the
         peak height of $1.3 \times 10^{-7}M$ Percaine Ciba in 0.4M KCl + 0.01M
         H_2SO_4. Curve 1, only supporting electrolyte accumulation
         time 360 s at -0.5 V (SCE); curve 2, with addition of 1.3 x
         $10^{-7}M$ Percaine, without accumulation; curve 3, same as curve
         2 but after accumulation (360 s at -0.5 V). Scan rate
         20 mV/s.

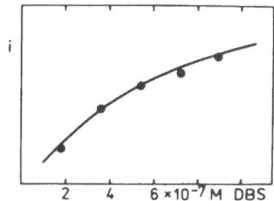

Fig. 4. The concentration dependence of the peak height of
         dodecylbenzensulphonate in 5M NaOH. Accumulation time
         180 s, unstirred.

Table 1. Conditions for DPP-adsorptive-stripping Analysis. Duration
          of Accumulation 180s to 360s (unstirred); Electrode area
          0.042 cm^2.

Compound	Electrolyte	Concentration M	Accumulation potential V(SCE)	Peak potential V(SCE)
codeine	1M NaOH	10^{-6}	-0.7	-1.1
	5M NaOH	5.10^{-8}	-1.0	-1.2
papaverine	1M NaOH	10^{-7}	-0.7	-1.4
atropine	0.05M NaOH	10^{-6}	-0.7	1.3
cocaine	0.05M NaOH	10^{-6}	-0.7	1.2
laurylsulphonate	1M NaOH	10^{-5}	-0.9	-1.3
dodecylbenzen-sulphate NA	5M NaOH	10^{-7}	-1.0	-1.2
Intercaine (p-Butylaminobenzoxyl-dimathylamino-ethanol hydro-chloride)	0.05M NaOH	10^{-7}	-0.7	-1.2
Percaine (2-Butoxy-4-diethyl-amino-ethylamino-carboxyquinoline hydrochloride)	0.4M NaCl 0.01M H$_2$SO$_4$	10^{-8}	-0.5	-0.75
Ge(IV)	0.1M H$_2$SO$_4$, 0.15 pyrocatechine	10^{-8}	-0.3	-0.55
Cu(II)	0.25M NH$_4$CNS 0.25 H$_2$SO$_4$	10^{-8}	-0.3	-0.6

In connection with the peak potential, it should be mentioned
that, as a rule, with increasing amount of the adsorbed compound at
the electrode surface, the measured desorption peak is shifted to
more negative potential values.

The accumulation effect can be exploited also for adsorptive
stripping analysis of polarographically reducible surface active
compounds. As an example, the local anaesthetic Percaine Ciba (Table
1) can be mentioned; the obtained peak belongs to the reduction of
the adsorbed compound (Figure 3). Other examples of adsorptive
accumulation of electroactive compounds are the Ge (Figure 5)
and Cu surface active complexes mentioned in Table 1. In the case of
the adsorbable complexes mentioned the method is suitable mainly for
metals where the amalgam formation is complicated or where the solubility of the metal in mercury is low.

The adsorptive accumulation of organic compounds described above
can be used also in combination with liquid chromatography. In these
measurements the Hg electrode in the detector unit is held at the
accumulation potential (this potential does not differ very much for
compounds of similar structure). The accumulation procedure is
periodically interrupted by a fast voltage scan (up to 100 mV/s scan
rate) to more negative potentials and the polarographic curve
recorded. Thus a set of polarograms of the dependence of the column
efluent volume is obtained. Of course other modes of measurement are
also possible.

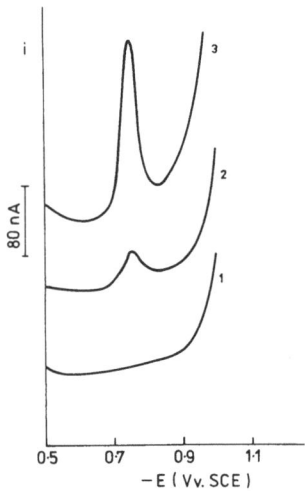

Fig. 5.   Stripping DPP curves of 2 x 20^{-8}M GeO$_2$ in 0.1M H$_2$SO$_4$ + 0.15M
          pyrocatechine after adsorptive accumulation. Unstirred.
          The duration of accumulation at -0.3 V(SCE): Curve 1, 0;
          curve 2, 360 s; curve 3, 720 s; curve 4, 1800 s.

SUMMARY

The adsorptive accumulation of some alkaloids, local anaesthetics, surfactants and metal complexes in stripping analysis using differential pulse polarographic procedure is described. The combination of this stripping method with liquid chromatography is possible.

REFERENCES

1. L. Novotný, The multipurpose dropping mercury electrode, Proc. J. Heyrovský Memorial Congress on Polarography, Vol.II, p.129, Prague (1980).
2. A. Y. Gokhshstein and Y. P. Gokhshstein, Zhur.Fiz.Khim., 36:651 (1962).
3. W. M. Peterson, Internat.Laboratory, 51 Jan/Feb (1980).
4. R. Kalvoda, Techniques of Oscillographic Polarography, Elsevier, Amsterdam, (1965).
5. R. Kalvoda and G. K. Budnikov, Collection Czechoslov.Chem., Commun., 28:838 (1963). 6. R. Kalvoda, Anal.Chim.Acta, (Submitted for publication).

APPLICATIONS OF IMMISCIBLE ELECTROLYTE INTERFACES IN ANALYTICAL CHEMISTRY

Vladimír Mareček

J. Heyrovský Institute of Physical Chemistry and
Electrochemistry, Czechoslovak Academy of Sciences
U továren 254, 102 00 Prague 10 - Hostivař, Czechoslovakia

INTRODUCTION

The electrolysis at the interface between two immiscible electrolyte solutions (ITIES) has recently been developed as a new electroanalytical method[1]. It is based on monitoring the current signal which corresponds to the charge transfer across the water/organic solvent (e.g. nitrobenzene) interface. The charge transfer may be either simple ion or simple electron transfer as well as either of these reactions coupled to one or a series of chemical reactions occuring in the bulk of the phases in contact.

In the first instance ion transfers were investigated such as the transfer of quaternary ammonium[2-4] or cesium[5,6] cations and picrate[7], CNS^-, ClO^-_4, laurylsulfate and n-octoate[8] anions. These processes are quite rapid with an apparent rate constant which exceeds 10^{-2} cm s^{-1} [5,6]. The only example of electron transfer reaction which has been observed was between the hydrophobic ferrocinium - ferrocene redox couple in nitrobenzene and the hydrophilic hexacyanoferrate redox couple in water[9]. A more complex mechanism is involved in the case of ion transfer facilitated by an ionophore[10]. This is the case, for example in the transfer of the alkali and alkaline earth metal cations across a water/nitrobenzene interface facilitated by synthetic neutral cyclic or acyclic ionophores derived from 3,6-dioxaoctanedicarboxylic acid[11].

In this communication our attempts to exploit the electrolysis at ITIES for analytical purposes are described. As examples the determination of acetylcholine, tetraethylammonium, calcium, barium and strontium cations by differential pulse stripping voltammetry (DPSV) at the hanging electrolyte drop electrode (HEDE) will be presented.

Experimental

Currently a four-electrode system with automatic ohmic drop compensation[5] has been used for accurate polarization measurements at ITIES. However, this rather complicated experimental set-up and the large area of the water/nitrobenzene interface (about 100 mm^2)[5] have not permitted the use of the fast pulse technique. Therefore, we have developed a simpler three-electrode system with the hanging electrolyte drop electrode.[1]

Assembly of the HEDE, which is shown in Figure 1, has been described elsewhere.[1] The small drop of nitrobenzene solution was formed at the capillary tip of a glass tube which was immersed in the test aqueous solution. The capillary tip of about 1 mm i.d. was drawn from the branched 4 mm i.d. glass tube. The volume of the nitrobenzene drop was controlled by a calibrated microsyringe which was connected to one branch of the glass tube while in the other two branches the metal wire counter electrode CE 1 and the Ag/AgCl reference electrode RE 1 were fixed. The latter reference electrode was dipped into the aqueous solution of tetrabutylammonium or tetraphenylarsonium chloride according to which base electrolyte was used for the nitrobenzene phase. The counter electrode CE 1 was the platinum or copper wire insulated by Teflon so that only the metallic disc (area of 0.126 mm^2) was exposed to the nitrobenzene solution. A motor-driven glass stirrer and the Ag/AgCl reference electrode RE 2

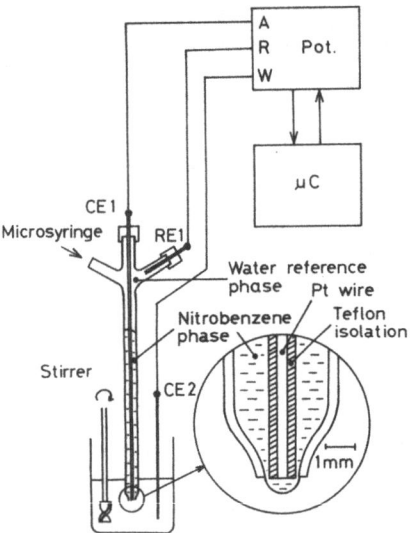

Fig. 1.   Assembly for hanging electrolyte drop electrode. CE 1 and
          CE 2 are the counter electrodes, RE 1 is the reference
          electrode, Pot. is the conventional three-electrode
          potentiostat with the IR drop compensation, μC is the
          microcomputer.

were dipped in the test aqueous solution. The counter electrode CE
1, the Ag/AgCl reference electrode RE 1 and the Ag/AgCl reference
electrode RE 2 were connected to the outputs for the auxiliary,
reference and working electrodes of the conventional three–electrode
potentiostat with positive feedback for an IR potential drop compensation. In this way, the metallic connection to the silver of the
reference electrode RE 2 was held at virtual ground while the potential E (vs. ground) at the metallic connection to the silver of the
reference electrode RE 1 was varied according to the chosen program.
The current was supplied from the output for the auxiliary electrode
by means of the counter electrode CE 1, it passed through the water/
nitrobenzene interface at the surface of the drop and it was picked
up by the reference electrode RE 2 which comprised the counter electrode CE 2 fo the four-electrode system.[5] The polarization of this
reference electrode by the current flowing through the system is
avoided in practice by using a large-area Ag/AgCl electrode.

With respect to the same concentrations of Cl^- ion in both
electrolytes, in which the Ag/AgCl electrodes are immersed, and with
respect to the same concentrations of the tetraphenylarsonium or
tetrabutylammonium ion in the nitrobenzene and aqueous phases of RE
1, the potential difference E is practically

$$E = \Delta_n^w \phi - \Delta_n^w \phi_{TA}^\theta{}^+ \quad + \quad I(R_w + R_n) \tag{1}$$

where $\Delta_n^w \phi = \phi(w) - \phi(n)$ is the Galvani potential difference between
the test aqueous and the nitrobenzene solutions; $\Delta_n^w \phi_{TA}^\theta{}^+$ is the formal
potential for TBA^+ or $TPAs^+$ cation ($\Delta_n^w \phi_{TBA}^\theta{}^+ = -0.248$ V; $\Delta_n^w \phi_{TPAs}^\theta{}^+ =$
-0.372 V) calculated from the extraction data,[12] and the last term
is the sum of the ohmic potential differences in the aqueous and
nitrobenzene solutions.

The program voltage for the potentiostat is generated by a 5 μs,
12 bit digital-to-analog converter with the 1 mV LSB controlled by a
microcomputer where the experimental current data are stored after
conversion to digital form by a data acquisition system. The data
acquisition system is composed of an external gain, programming-
instrumentation differential amplifier, 9 μs sample and hold unit and
24 μs, 12 bit analog to digital converter. Experimental data stored
in the memory can be rewritten on an X-Y recorder or a storage
oscilloscope. The whole system including number oriented microprocessor is controlled by the microprocessor Intel 8080.

The potential waveform for DPSV is shown in Figure 2. The
staircase voltage pulse 1 mV per 20 ms starts at the initial potential E_i with respect to the potential of preelectrolysis E_e the
duration of which was $t_e = 60$ s in all experiments. During this
period of time the glass stirrer in the test aqueous solution was
automatically switched on. In the determination of acetylcholine and
tetraethylammonium cations, the time intervals t_1 and t_2 were 100 ms.

The current was sampled and averaged during the last 20 ms of the
time intervals t_1, i.e., at the points C_1 and C_2 (Figure 2). The
difference of the averaged currents at C_1 and C_2 was stored in the
computer memory. In order to increase the sensitivity of DPSV for
the determination of Ca^{2+}, Ba^{2+} and Sr^{2+} cations the time interval
t_1 was decreased to 20 ms and the time interval t_2 was set to zero.
Moreover, the current was sampled only once at the end of the time
intervals t_1. To minimize the noise, the current readout was
synchronized with the line frequency 50 Hz. The difference of the
currents at C_1 and C_2 was again stored in the computer memory. The
pulse magnitude ΔE was typically 20 mV except for the determination
of TEA^+ cation in which case $\Delta E = 40$ mV.

DPSV of Quaternary Ammonium Cations

The determination of the quaternary ammonium cation $(CH_3)_3N^+CH_2$
$OCOCH_3$ commonly called acetylcholine[1] (ACH^+), which is extremely
important in controlling the function of the nervous system, is
illustrated in Figure 3. Below the initial potential $E_i = 250$ mV the
transfer of ACH^+ from water to nitrobenzene can hardly be detected
while above $E_e = 550$ mV this transfer is very fast and controlled by
the diffusion. The DPSV curve of acetylcholine, which has been
previously concentrated into the nitrobenzene drop, is the symmetric
peak with the peak potential $E_p = 420$ mV (Figure 3a). In Figure 3b
the peak height is plotted vs. concentration of acetylcholine chloride, which is expressed in part per million units (ppm), with 1 ppm
corresponding to 5.5 µM concentration of acetylcholine. Good linearity exists for the concentration of acetylcholine chloride in
water between 0.5 and 5.0 ppm. The peak potential E_p coincides with
the polarographic half-wave potential for acetylcholine derived from
cyclic voltammetry [13].

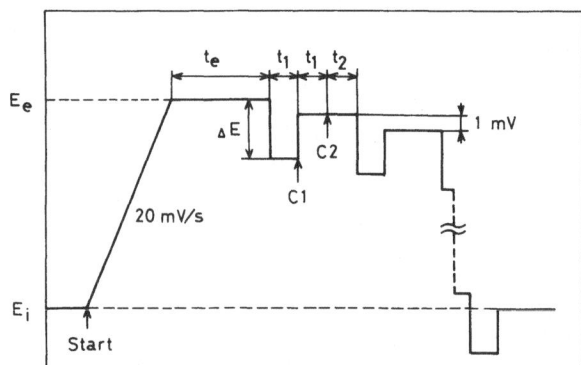

Fig. 2. Potential waveform for differential pulse stripping
         voltammetry. E_i - initial potential, E_e - potential of
         pre-electrolysis, $t_{1,2}$ - variable time intervals, $C_{1,2}$ -
         current sampling points.

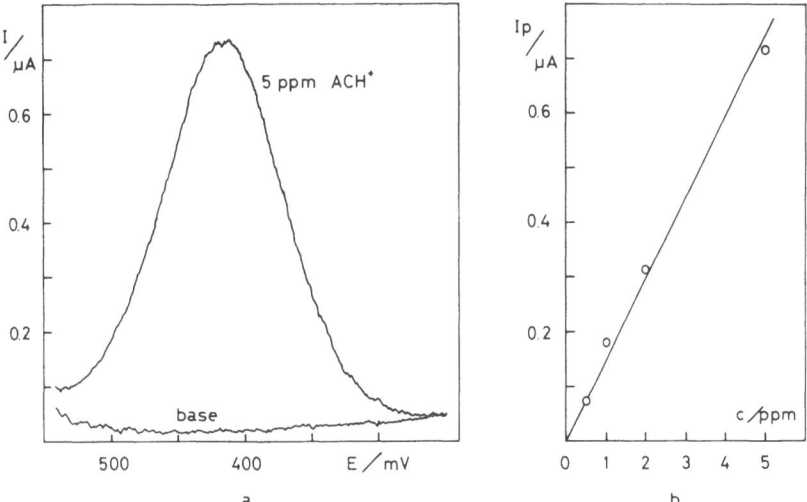

Fig. 3. (a) Differential pulse stripping voltammogram of 0.01 M LiCl
         in water and 0.01 M tetraphenylarsonium tetraphenylborate in
         nitrobenzene in the presence or absence of 5 ppm ACHCl in
         water. Ohmic drop compensation: 4.75 KΩ.
         (b) Plot of the peak height of acetylcholine vs. con-
         centration of ACHCl in water.

     The DPSV curve of another quaternary ammonium cation - tetra-
ethylammonium - has a similar character. In Figure 4 the peak height
of DPSV curve of TEA$^+$ is plotted vs. concentration of TEABr in water,
expressed in ppm with 1 ppm corresponding to 4.8 μM concentration of
TEA$^+$. If the potential drop due to the solution resistance, 4.75 kΩ,
is compensated, good linearity exists for the concentrations of TEABr
between 0.5 and 10 ppm (plot a in Figure 4). Without the ohmic drop
compensation there is some deviation from linearity (plot b in Figure
4) for concentrations higher than 5 ppm, reaching 20% at 10 ppm.
This deviation is in agreement with the rough estimate. Thus, for
the electrolyte resistance 4.75 kΩ the ohmic potential drop should be
about 10 mV at the peak potential for 10 ppm concentration of
TEABr [14]. Because the peak current is proportional to the pulse
magnitude, ΔE, its decrease by δΔE ≐ IR ≐ 10 mV should yield a
decrease in current of about 25%.

DPSV of Ca^{2+}, Ba^{2+} and Sr^{2+} Cations

     The determination of calcium, barium and strontium cations is an
example of the ionophore (X) facilitated charge transfer process

$$M^{Z+}(W) + X(n) = MX^{Z+}(n) \qquad\qquad (2)$$

controlled by the transport of the metal cation in the aqueous phase,
i.e. when $[M^{2+}(w)] \ll [X(n)]$. As an ionophore a cyclic polyetherdiamide
(7,19-dibenzyl-2,3-dimethyl-7,19-diaza-1,4,10,13,16-pentaoxaheneico-
sane-6,20-dione)[15] was used.   Table 1 collects the stability constants K of complexes in nitrobenzene for several cations obtained by
cyclic voltammetry[11] together with the standard electrical poten-
tial difference $\Delta_n^w \phi^\circ$ for the ion transfer from water to nitrobenzene.
Even though the stability constants for calcium, barium and strontium
cations in nitrobenzene are much higher than that for lithium cation,
the translocation of lithium ion from water to nitrobenzene is
shifted by about 300 mV negatively when it is used as a base electrolyte, i.e., when $[Li^+(w)] \gg [X(n)]$ and it coincides with the transfer
of alkaline earth metal cation.   From a comparison of the standard
electrical potential difference and from the calculated half-wave
potentials using the stability constant for several cations we have
chosen magnesium chloride as a base electrolyte for the aqueous
phase.   To minimize the base electrolytes' current in DPSV we have
decreased their concentrations.   In the case of magnesium chloride it
was 2.5 mM solution in water and 5 mM solution of TBATPB in nitrobenzene.   The concentration of the ionophore was 1 mM.

The DPSV experimental curves of Ba^{2+}, Sr^{2+} and Ca^{2+} cations are
compared in Figure 5a.   The initial potential E_i for each cation was
180 mV at which the transfer of these ions from water to nitrobenzene
is negligible.   The potential of pre-electrolysis of Ba^{2+} and Ca^{2+}
was E_e = 380 mV, while for Sr^{2+} E_e = 400 mV.   The concentration of
the ion was 4 µM in each case.   The DPSV peak potential in the case
of Ba^{2+} cation was 315 mV which agrees perfectly with the theoretical
half-wave potential 305 mV (Table 1).   Because the Ca^{2+} transfer is

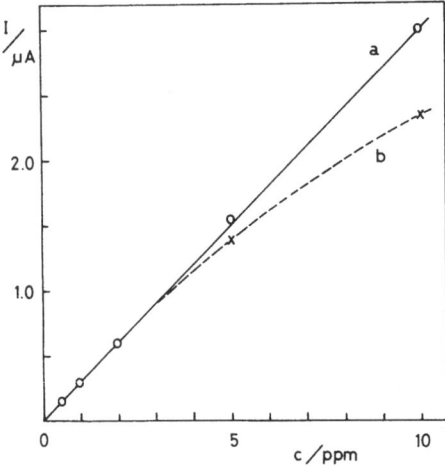

Fig. 4.   Plot of the peak height of TEA$^+$ vs. concentration of TEABr
in water with (a) or without (b) the IR drop compensation,
R = 4.75 kΩ.

not fast enough, as is the case of Ba^{2+} transfer, the DPSV current peak, which approximately correspond to the first derivative of the normal polarographic wave, is smaller than in the case of Ba^{2+}. The peak potential is 262 mV which again agrees with the theoretical half-wave potential 254 mV. From a comparison of the DPSV peak currents of calcium and strontium cations it can be seen that the facilitated transfer reaction of Sr^{2+} cation is more rapid than that of Ca^{2+} cation. In Figure 5b the peak height vs. concentration for Ca^{2+}, Ba^{2+} and Sr^{2+} is plotted. Good linearity is observed for concentrations between 1 and 4 ppm.

Table 1

		Li^+	Mg^{2+}	Ca^{2+}	Sr^{2+}	Ba^{2+}
$\Delta_n^w \phi^\theta$ (mV)		395	361	349	342	341
$\log K /M^{-1}$ or M^{-2}		7.19	11.96	18.30	16.91	16.30
$\Delta \phi 1/2$ (mV)	$[M^{2+}(w)] \gg [X(n)]$	106	190	− 6	27	47
	$[M^{2+}(w)] \ll [X(n)]$	176	203	6	41	57

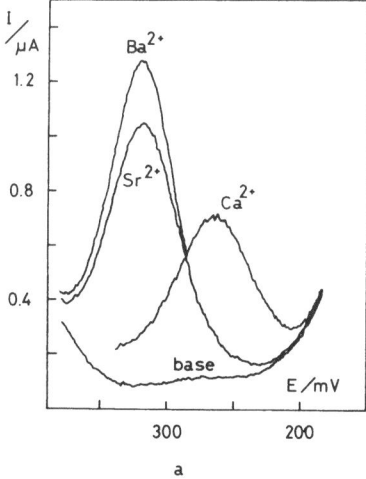

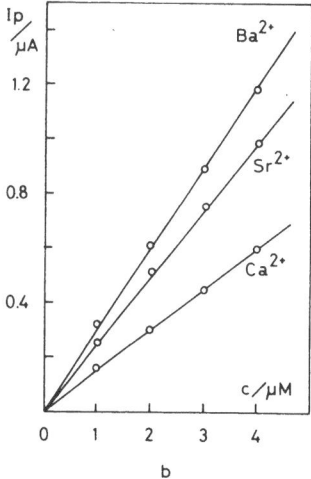

Fig. 5. (a) Differential pulse stripping voltammograms of 4 μM solutions of Ca^{2+}, Ba^{2+} and Sr^{2+} cations in water. Base electrolytes: 2.5 mM $MgCl_2$ in water, 5 mM tetrabutyl-ammonium tetraphenylborate and 1 mM ionophore (see text) in nitrobenzene. Ohmic drop compensation: 6 kΩ.
(b) Plots of the peak height of Ca^{2+}, Ba^{2+} and Sr^{2+} cations vs. their concentration in water.

It can be concluded that electrolysis at the interface between two immiscible electrolyte solutions offers quite attractive opportunities for exploitation in chemical analysis. The first attempts are promising and future research hopefully will reveal more of the charge transfer reactions of analytical interest.

REFERENCES

1. V. Marecek and Z. Samec, Anal.Lett., A14 (1981).
2. Z. Samec, V. Marecek, J. Weber, and D. Homolka, J.Electroanal. Chem., 99:385 (1979).
3. C. Gavach and F. Henry, J.Electroanal.Chem., 54:361 (1974).
4. C. Gavach and B. d'Epenoux, J.Electroanal.Chem., 55:59 (1974).
5. Z. Samec, V. Marecek, and J. Weber, J.Electroanal.Chem., 100:841 (1979).
6. Z. Samec, V. Marecek, J. Weber, and D. Homolka, J.Electroanal. Chem., 126:105 (1981).
7. D. Homolka, and V. Marecek, J.Electroanal.Chem., 112:91 (1980).
8. P. Vanýsek, J.Electroanal.Chem., 121:149 (1981).
9. Z. Samec, V. Marecek, and J. Weber, J.Electroanal.Chem., 103:11 (1979).
10. J. Koryta, Electrochim.Acta, 24:293 (1979).
11. Z. Samec, V. Marecek, and D. Homolka, J.Electroanal.Chem., submitted for publication.
12. J. Rais, Collect.Czech.Chem.Commun., 36:3253 (1971).
13. P. Vanýsek, and M. Behrendt, J.Electroanal.Chem., submitted for publication.
14. V. Marecek, and Z. Samec, Anal.Chem., submitted for publication.
15. J. Petránek, and O. Ryba, Anal.Chim.Acta, 128:129 (1981).

THE IMPEDANCE OF SMALL Li-CuO PRIMARY CELLS

R. Leek, S.A.G.R. Karunathilaka,* N.A. Hampson**
and T. J. Sinclair***

*Electronic and Electrical Engineering Department
Loughborough University of Technology
Loughborough, LE11 3TU UK
**Chemistry Department, Loughborough University of
Technology, Loughborough, LE11 3 TU UK
***Procurement Executive, Ministry of Defence, Royal
Armament Research and Development Establishment,
Fort Halstead, Sevenoaks, TN14 7BP UK

SUMMARY

The impedance of small lithium-copper oxide primary cells has
been investigated in a frequency range from 5 mHz to 10 kHz. The
cells had been stored after assembly for from three weeks up to three
years and their state of charge was from 100% down to 20%. After an
initial period of electrochemical stabilization, the cells exhibited
consistent results and the shape of the impedance locus was found to
depend markedly on the state of charge of the cell. An interpretation of the results is given in terms of an analogue circuit which
contains components to represent the contribution to the impedance of
each electrode and of the electrolyte.

INTRODUCTION

The lithium-copper oxide cells used in this investigation were
made by SAFT and had a type specification LC01. Each cell has a
negative electrode of lithium metal, an electrolyte of lithium perchlorate in an organic solvent and a positive electrode of copper
oxide and carbon. The cell is constructed in the form of a hollow
tubular lithium central cylinder which is separated by a thin annular
synthetic material from a concentric tubular copper oxide-carbon ring
matrix enclosed by an outer steel can. Current collectors are at-
tached to the lithium and steel cylinders.

149

Initially the central cavity of the cell is filled with electrolyte solution that passes by capillary action into the copper oxide-carbon matrix which swells to form a spongy mass. After a time, all the electrolyte is removed from the central cavity and becomes distributed homogeneously throughout the space between the lithium and the steel cylinders. After an initial period of electrochemical stabilization, very little further change occurs and the cell is available for use.

Lithium-based primary cells are being increasingly used for high specific energy applications in which low current density discharging only is demanded. Specific energies of about 700 W h dm^{-3} are not uncommon. Another feature of the lithium non-aqueous system is the very long times which can be tolerated in the inactive condition without deterioration of the stored electrical capacity. Storage times in excess of ten years have been claimed in the case of the Li-CuO system.

The chemistry of Li-CuO cells is not completely understood. In principle the mechanism

$$2Li + CuO \rightarrow Li_2O + Cu \qquad (1)$$

is far too simple. Other possibilities can be proposed based on the formation of Li_2O_2 and Cu_2O; however, none of these is really satisfactory.

There may be intercalation of lithium species within the CuO-C electrode which ultimately becomes transformed to a mixture of copper and Li_2O; alternatively a mixture of Cu_2O and Li_2O is produced. Intermediate compounds of the type CuOLi have been suggested; however, the mechanism for the formation of such species seems obscure. A further factor which is as yet not understood is that the single-step discharge curve (1.5 V) at 20°C becomes a two-step curve (1.6 V and 1.4 V) at 70°C similar to that of AgO electrode.

As part of a program aimed at the prediction of the residual capacity of primary cells the impedances of small Li-CuO cells have been measured over a large frequency range. The results are recorded and an interpretation of the data is given in this paper.

Experimental Details

The experimental method and apparatus have been fully described elsewhere [1,2]. The cells used were manufactured by SAFT Sogea and were type LC01 which is the R6 or AA size. The cells, which were 13.85 mm in diameter and 49.5 mm long with a volume of 7.45 cm^3 and a mass of 17.4 g, had a nominally rated energy density of 200 W h kg^{-1} when discharging into 120 Ω at 20°C with a mean voltage of 1.5 V.

The actual surface area of the lithium electrode was 9.954 cm^2 and
the electrolyte solution filled a volume of 2.35 cm^3. The impedances
of newly received and aged cells were measured, and the cells were
then discharged galvanostatically at the 100 h rate (37.5 mA) using a
Kemitron P-50 potentiostat in a galvanostatic configuration with a
CM-2 coulometer. Impedance measurements were made at a series of
discharge states.

Results

Figure 1 shows the impedance locus characteristic of the cell as
received from the manufacturer, i.e. after assembly, filling and a
period of storage during which time the electrolyte (introduced into
the hollow center of the cell) is distributed uniformly throughout
the cell. The electrolyte solution is absorbed to a large extent
into the CuO-C positive electrode which becomes quite swollen. If
the initial storage period is not sufficiently long the impedance
changes until equilibrium is obtained. Once the cell is equilibrated
little further change occurs. Figure 2 shows the impedance locus of
the cell after open-circuit storage for a further three weeks; the
similarity to Figure 1 is marked, and the differences that do exist
are only concerned with magnitude. Figure 3 shows the impedance
spectrum of a cell after storage for three years. Here again the
shape of the locus is unchanged, the magnitude differences being
relatively minor between the cells stored for six weeks and those
stored for three years. The loci shown in Figures 1-3 are interesting for they show an elongated high frequency semicircle corresponding to a charge transfer process; the linear section is at an angle
of approximately 45° to the horizontal and this is maintained down to
about 0.5 Hz when the slope begins to increase markedly.

On discharging, the magnitude of the locus changes. Figure 4
shows the effect of a 1% removal of capacity. The high frequency
semicircle has shrunk indicating a more rapid charge transfer process
and the degenerate line following the semicircle leaves the real axis
at an angle more nearly 45° than that obtained for the undischarged
cell. The increase in slope at the lower frequency still remains a
well-developed feature.

Continual discharge to 5% gives rise to a much more complicated
locus. It should be emphasized that discharging produces some time
instability of both open-circuit potential and open-circuit impedance. Figures 5-11 show the impedances after various times on
open circuit following discharge to the 5% level. The potential
slowly increases over 120 h from about 1.8 V to 1.94 V, and the
impedance locus, although maintaining the same general shape, changes
somewhat in magnitude. This tendency for the cells to require time
for stabilization has to be taken into account in all our investi-
gational work and consequently adequate time has to be allowed be-

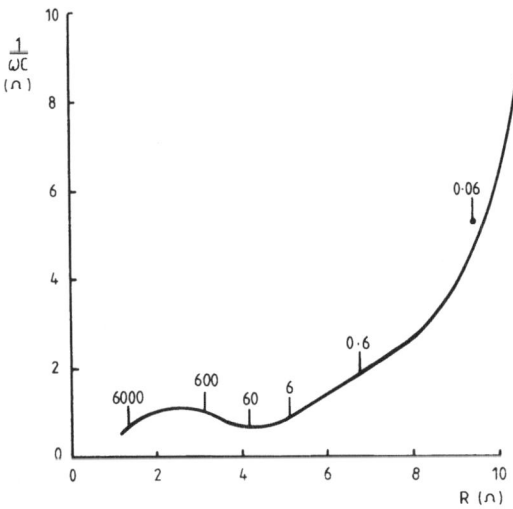

Fig. 1.   Locus of the impedance 3 weeks after assembly (open–circuit voltage, 2.38V).

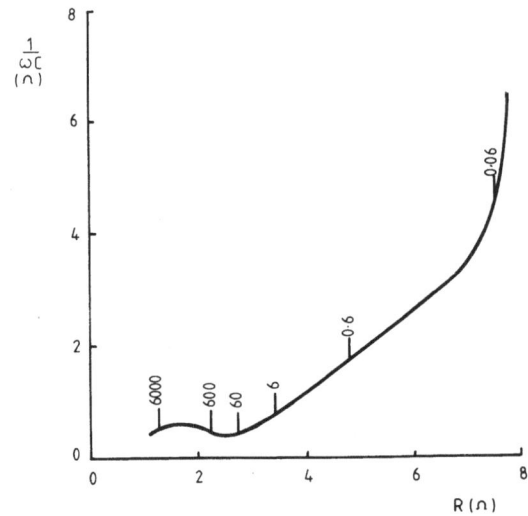

Fig. 2.   As for Fig. 1 but 6 weeks after assembly (open–circuit voltage, 2.38V).

tween the end of a discharging period and the start of an impedance experiment. Generally about 24 h is required to give an impedance locus which experience indicates is sufficiently stable for the impedance data to be representative of the state of charge. Figures 5–11 for the 5% discharged state show indications of a second semi-

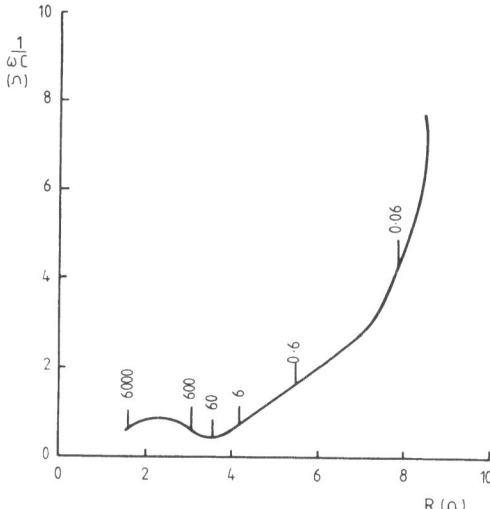

Fig. 3.   As for Fig. 1 but 3 years after assembly (open-circuit voltage, 2.20 V).

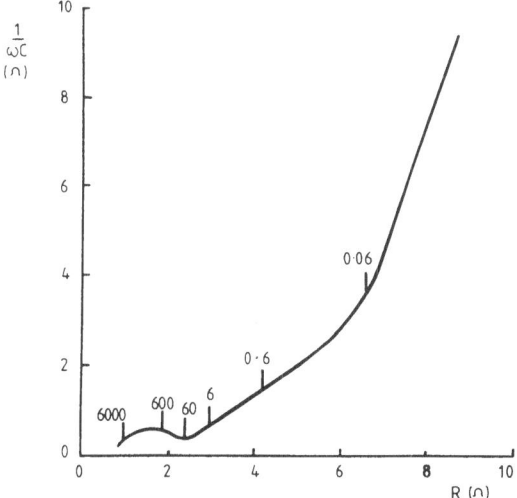

Fig. 4.   Impedance locus of a 1% discharged cell after 72 h at the 100 h rate (37.5 mA) (open-circuit voltage, 2.09 V).

circle on the low frequency side of the first (high frequency) semi-circle. This becomes more evident at the 10% discharged state (Figure 12), and is very well defined at the 20% and the 30% dis-charged states as shown in Figures 13 and 14. It should further be noted that, on discharging, the linear region at 45°, initially

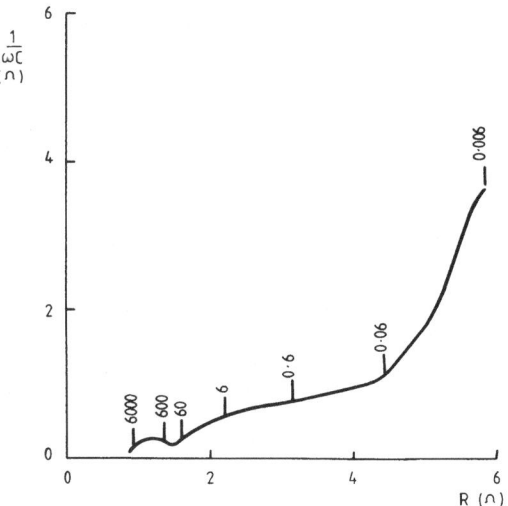

Fig. 5.   As for Fig. 4 but 5% discharged with 4 h rest (open-circuit voltage, 1.809 V).

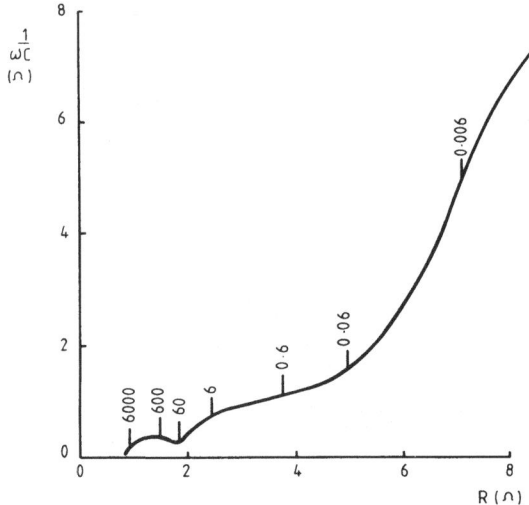

Fig. 6.   As for Fig. 5 but with 19 h rest (open-circuit voltage, 1.894 V).

observed immediately after the high frequency semicircle at high states of charge, is lost; the impedance locus ultimately progressively increases in slope as the frequency is reduced. Further discharging results in plots similar in form to Figure 13, as shown in Figures 14-19 as the capacity is drained down to 80% discharged.

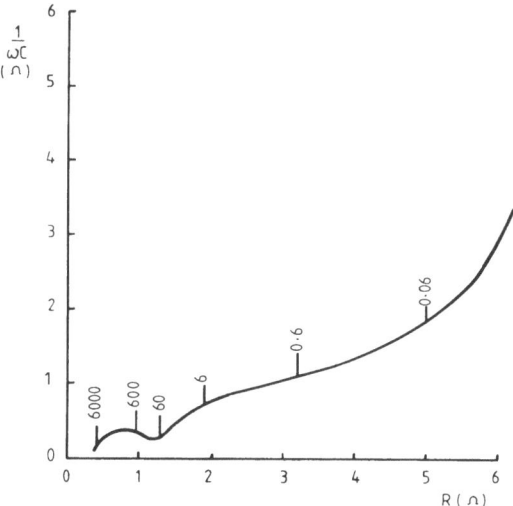

Fig. 7.   As for Fig. 5 but with 25 h rest (open-circuit voltage,
          1.903 V).

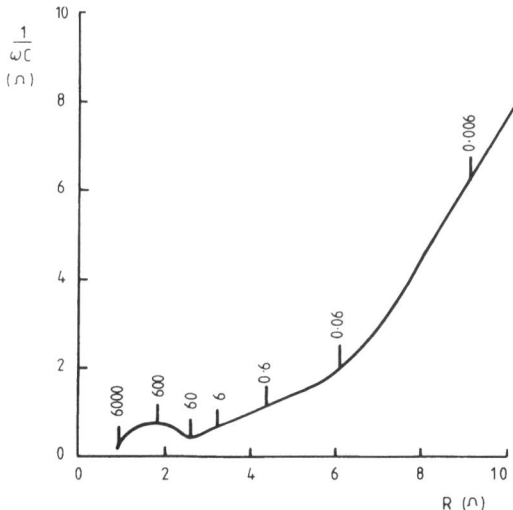

Fig. 8.   As for Fig. 5 but with 46 h rest (open-circuit voltage,
          1.908 V).

     It should be emphasized that the loci shown all pertain to cells
that had been stored for about 24 h between the end of discharge and
the impedance determination. This was necessary owing to changes
which occurred in the frequency response in the period immediately
following a discharge. These changes were very subjective and de-

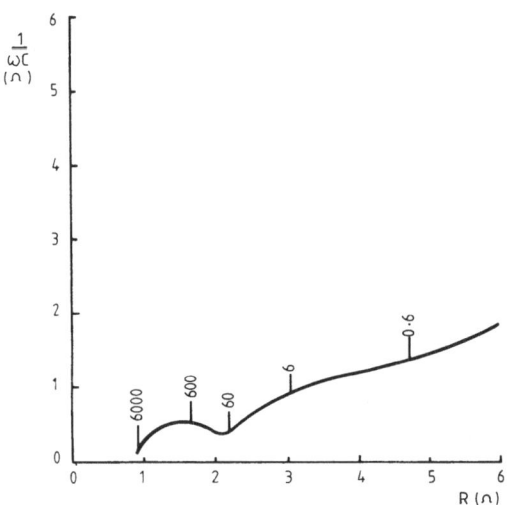

Fig. 9.   As for Fig. 5 but with 92 h rest (open-circuit voltage,
           1.936 V).

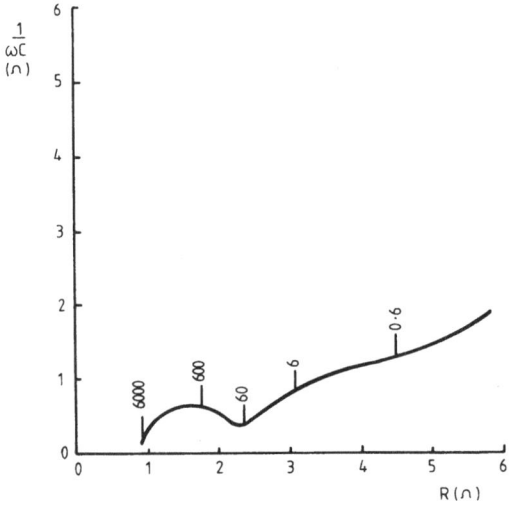

Fig. 10.   As for Fig. 5 but with 119 h rest (open circuit voltage,
           1.942 V).

pended strongly on the rate of discharging, the extent of the capacity withdrawn from the cell and the time elapsed between the end
of the discharge and the impedance experiment. The form of the
impedance change with time was always the same; the electrochemistry
became more kinetically hindered. After 24 h very little further
change was found to occur.

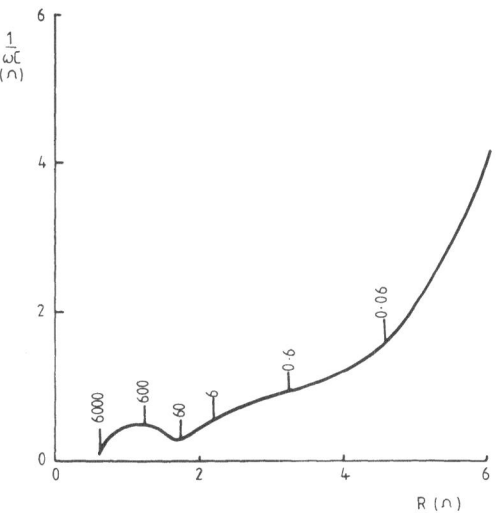

Fig. 11.   As for Fig. 5 but with 164 h rest (open–circuit voltage, 1.941 V).

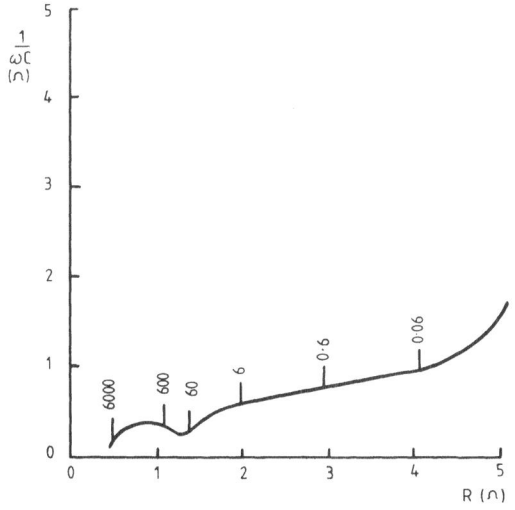

Fig. 12.   As for Fig. 4 but 10% discharged and with 19 h rest (open–circuit voltage, 1.601 V).

DISCUSSION

The cell consists of two electrodes and an electrolyte; consequently the simplest analogue is that of two Randles impedances[3] in series with the electrolye resistance as shown in Figure 20. The impedances of two-terminal cells have been discussed elsewhere[4] for

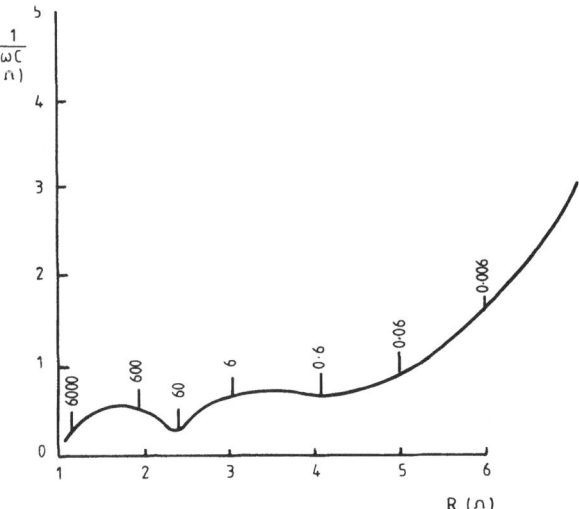

Fig. 13. As for Fig. 4 but 20% discharged and with 10 days rest
(open-circuit voltage, 1.662 V).

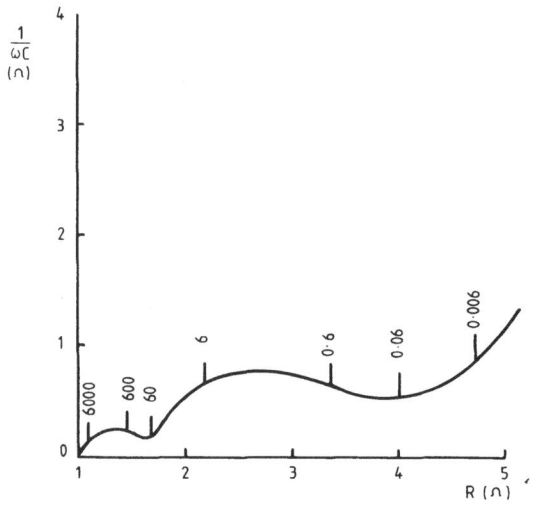

Fig. 14. As for Fig. 4 but 30% discharged and with 5 days rest
(open-circuit voltage, 1.604 V).

the case when the contribution of each electrode to the total impedance is about the same. The loci corresponding to these cases
show two semicircles with a Warburg region at low frequencies. If
the charge transfer resistances of the two electrodes are about equal
in magnitude then a single semicircle would be observed in the high

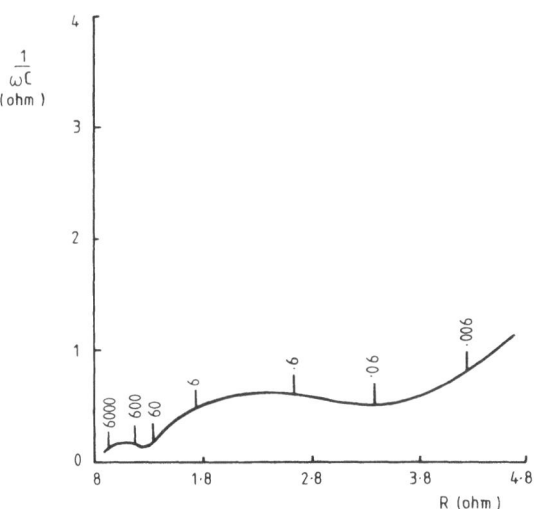

Fig. 15.   As for Fig. 4 but 40% discharged and with 7 days rest
          (open-circuit voltage, 1.797 V).

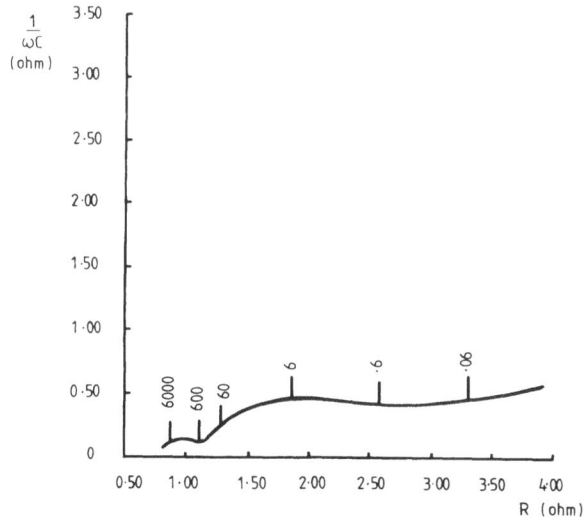

Fig. 16.   As for Fig. 4 but 50% discharged and with 5 days rest
          (open-circuit voltage, 1.29 V).

frequency region similar to that shown in Figure 3 for the cell
stored for 3 years. Distortion of the semicircle results when the
charge transfer resistance at one electrode is sufficiently different
from that at the other. This state of affairs could give rise to the
high frequency regions of Figures 1 and 2 which correspond to relatively newly assembled cells.

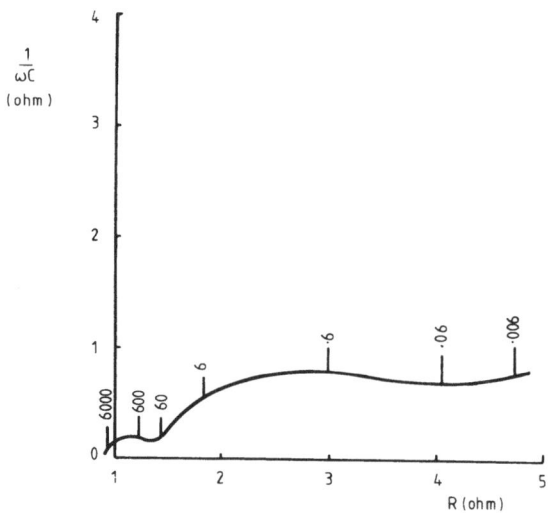

Fig. 17. As for Fig. 4 but 60% discharged and with 5 days rest
(open-circuit voltage, 1.715 V).

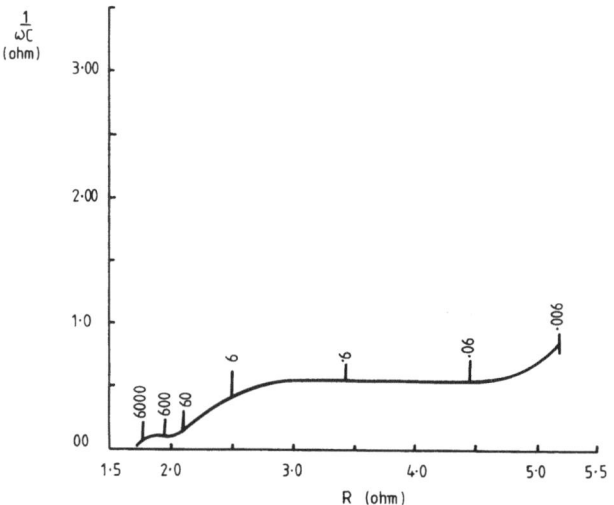

Fig. 18. As for Fig. 4 but 70% discharged.

The change in the dihedral angle of the Warburg tail at low
frequencies is a very interesting feature. This behavior has previously been encountered by Hampson et al.[5] in connection with PbO_2
electrodes and was interpreted as arising from the presence of a
layer of $PbSO_4$ of high dielectric constant at the front of the electrode. The effect on the cell analogue is to introduce an additional
series capacitor as shown in parentheses in Figure 20. At high

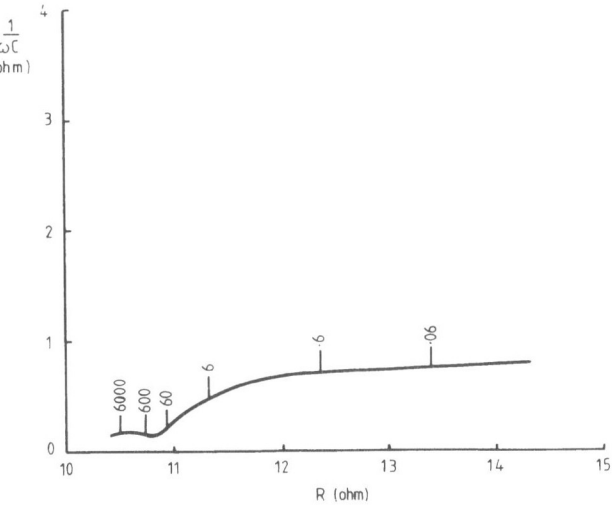

Fig. 19. As for Fig. 4 but 80% discharged.

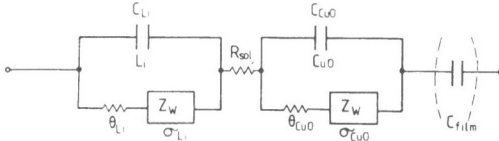

Fig. 20. Analogue circuit for the cell. The capacitor in
parentheses is evident at low frequencies.

frequencies the effect is minimal; however, at low frequencies it
results in a large reactive component.

Figure 21 shows the results of a computer simulation of the
frequency response of the circuit of Figure 20 without series capacitance. It is clear that, although the shape is generally satisfactory, the fine structure of the experimental loci for the new cell
has not been adequately reproduced, particularly in the sensitive
range where charge transfer control gives way to diffusion control,
i.e. at the junction of the semicircle and the Warburg region. It
should also be noted that it was necessary to include a roughness
factor in order to match the Warburg contributions correctly [2].

Figure 22 shows how the inclusion of a series capacitance ad-
justs the "normal" Warburg shape to fit the present system. We can
conclude at this point that such a model will adequately explain the
cell behavior.

The same general features are apparent for the discharged cells,
and consequently we can conclude that the same behavior resulting in
the same analogues and technical circuits applies. Figure 23 shows

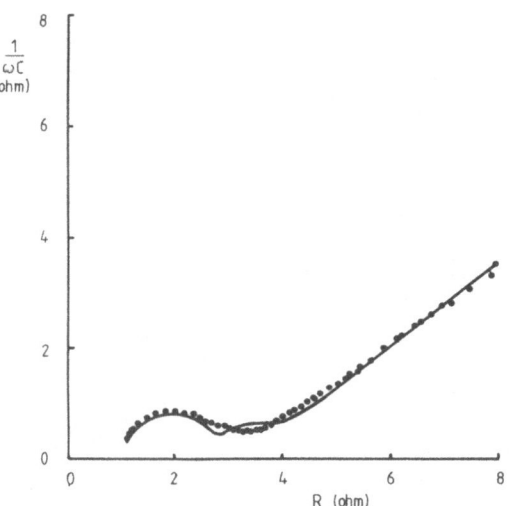

Fig. 21. Theoretical frequency response of the circuit of Fig. 20,
          series capacitor omitted.

$(C_{Li} = 50\mu F, \theta_{Li} = 1.7 , \sigma_{Li} = 3.5,$
$\gamma_{Li} = 0.7, C_{CuO} = 5.10^3\mu F, \mu F, \theta_{CuO} = 0.7 ,$
$\sigma_{CuO} = 1.0, \gamma_{CuO} = 1.0)$

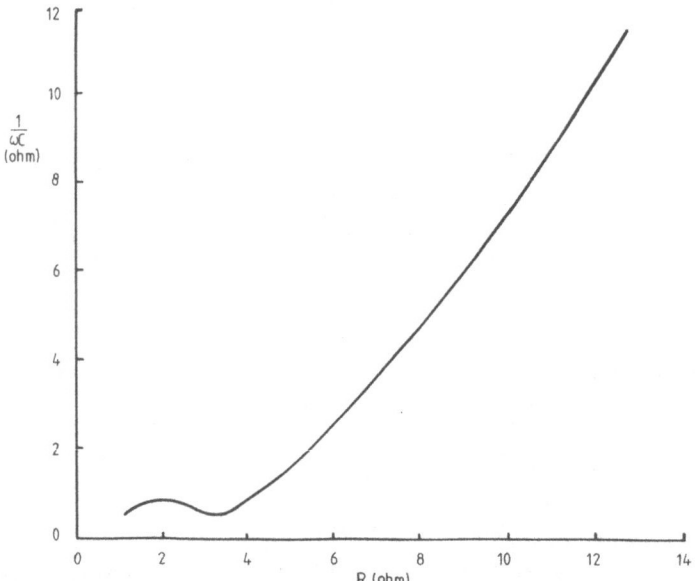

Fig. 22. The effect of series capacitance on the locus of the
          impedance of the circuit of Fig. 20.

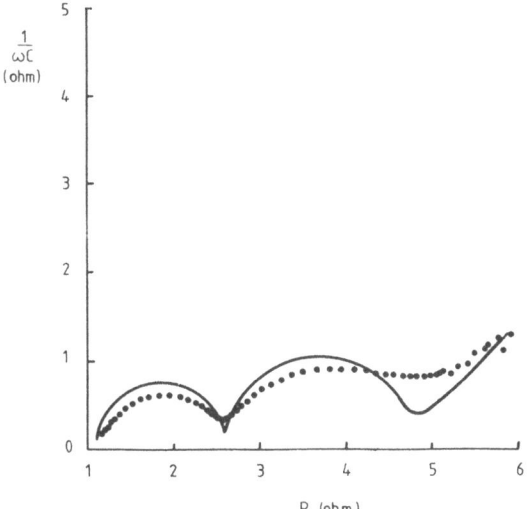

Fig. 23.   Theoretical frequency response of the circuit of Fig. 20
           and points obtained experimentally for a 20% discharged
           cell.

           (C_{Li} = 150µF, θ_{Li} = 1.5Ω, σ_{Li} = 0.1,

           γ_{Li} = 1.0, C_{CuO} = 35.7.10^3 µF, θ_{CuO} = 2.0,

           σ_{CuO} = 0.15, γ_{CuO} = 1.0, R_{Sol} = 1.1)

how computer simulation deals with matching the theoretical model to
a typical discharged cell. We have chosen a 20% discharged cell as
an example. It is clear from Figure 23 that the matching is not
particularly good. In fact the overestimates of the magnitudes of
the two distorted semicircles at high frequency suggests that the
complicated model of the partially discharged cell shown in Figure 20
is still too simple. It is probable that we have underestimated the
importance of the chemical reaction at the CuO electrode. It is
known that the electrochemistry involves more than the simple charge
transfer and a subsequent chemical reaction has been postulated. We
have already discussed such mechanisms theoretically[4], and our
calculations show that such a flattening of the semicircle would
occur as the result of the incorporation of a parallel resistance-
capacitance subcircuit element in series with the Warburg impedance
for the CuO electrode.

     We now intend to solve this model thoroughly and abstract the
individual values for the circuit elements using the Taylor expansion
technique employed in earlier work.

     Nevertheless, it is certain that sufficient change occurs during
discharging of the cell to form a suitable test of the state of
charge.

CONCLUSIONS

(1) The impedance of Li–CuO cells is complicated and the effects of
     both electrodes must be considered in modelling the cell
     analogue.
(2) The impedance changes markedly with the state of charge.
(3) Such changes will provide the necessary data for a test of the
     state of charge.

Acknowledgements

        We thank Procurement Executive of the Ministry of Defence for
financial support to S.A.G.R.K.

REFERENCES

1. N. A. Hampson and M. J. Willars, Surf.Technol., 7:247 (1978).
2. S. A. G. R. Karunathilaka, N. A. Hampson, R. Leek, and T. J.,
    Sinclair, J.Appl.Electrochem., 10:357 (1980).
3. J. E. B. Randles, Discuss.Faraday Soc., 1:11 (1947).
4. S. A. G. R. Karunathilaka, N. A. Hampson, and R. Leek, Surf.
    Technol., 13:339 (1981).
5. N. A. Hampson, S. Kelly, and K. Peters, J.Appl.Electrochem., in
    the press.

INDEX